MATHEMATICAL APPROACH TO PATTERN AND FORM IN PLANT GROWTH

MATHEMATICAL APPROACH TO PATTERN AND FORM IN PLANT GROWTH

ROGER V. JEAN
University of Quebec at Rimouski

WITHDRAWN

A Wiley-Interscience Publication
JOHN WILEY & SONS
New York · Chichester · Brisbane · Toronto · Singapore

Library of Congress Cataloging in Publication Data

Jean, Roger V.
    Mathematical approach to pattern and form in plant growth.

    Translation of: Croissance végétale et morphogénése.
    "A Wiley-Interscience publication."
    Includes bibliographical references and indexes.
    1. Phyllotaxis—Mathematical models.   2. Plant morphogenesis—Mathematical models.   I. Title.

QK649.J4313   1984   581.3   83-21875
ISBN 0-471-88357-3

Printed in the United States of America

10 9 8 7 6 5 4 3 2 1

à
Irène et Paul-Émile

FOREWORD

The patterns and forms of organisms are their most immediate and obvious properties. Accordingly, the study of pattern and form is the earliest aspect of biology. At the descriptive level, among many other things, it provided the foundation for the systematic taxonomy of Linnaeus, which is one of the cornerstones of biology. Like all good descriptive classifications (the periodic table in chemistry is another example) the Linnaean taxonomy manifests deep underlying organizational principles; this can be seen from the fact that taxonomy provided one of the basic clues which led to the theory of evolution.

The study of pattern and form was also central to the idea of the gene as an atomic unit of heredity. Mendel clearly regarded genes as units of morphological control. Indeed, the early development of genetics as a science, both in theory and in experiment, was tacitly predicated on the view that the role of genes was to control gross features of form in a whole multicellular organism.

One of the basic problems raised by the descriptive study of biological form is to understand how these forms are generated, and how the generation of form is controlled. This kind of question can be raised at two distinct levels. On the one hand, organic form is shaped by evolutionary processes; in this sense, the study of form is a phylogenetic question, to be attacked by (a) the study of natural selection operating on phenotypes, and (b) the effects of such phenotypic selection on the character of the underlying genotypes. On the other hand, the study of pattern and form is an ontogenetic question, which concerns the development of individual organisms from their earliest antecedents. There are many tantalizing relations between these two aspects of form generation, articulated among other ways in the "recapitulation" hypothesis of Haeckel ("ontogeny recapitulates phylogeny"), and in the transformation theory of D'Arcy Thompson.

The study of morphogenesis, which deals with the generation, stability, and control of pattern and form, is one of the most active areas of biology. At the same time, it remains one of the most refractory. The current explosion in molecular biology, far from illuminating the problems of morphogenesis, has only served to heighten their mystery. For instance, the Mendelian gene, initially conceived as a regulator of morphology in the intact organism, is now viewed exclusively as a modulator of intracellular chemistry. The resulting schism between chemistry and geometry has widened even further the gap between what we know and what needs to be explained.

One of the most intriguing and natural arenas for the study of morphogenesis, in its widest sense, is phyllotaxis. Classically, phyllotaxis refers to the positioning of the leaves of higher plants around an axis of growth; thus systematics of higher plants rests to a large extent upon phyllotaxis. But it is much more than that. Descriptive phyllotaxis is intimately connected with certain classical problems in number theory, especially with the well-known Fibonacci numbers. The relation between phyllotaxis and Fibonacci sequences is clearly not an accidental or random one, and points to some deep mechanism of morphogenetic control, which is manifested in ontogenesis and strongly conserved in phylogenesis. Indeed, these same Fibonacci sequences arise in the branching patterns of even the most primitive plants.

Thus the phenomena of phyllotaxis provide a natural laboratory in which to study general phenomena of morphogenesis. Understanding phyllotaxis may well provide the key for unlocking the ancient mysteries surrounding the generation of pattern and form in the organic world.

Dr. Jean has undertaken this difficult task. In this volume, he reviews the phenomena of phyllotaxis at the descriptive level and sets forth a number of competing approaches (including his own) that seek to explain them. Dr. Jean's own approach has, I believe, served to inject some essential new ideas into the study of phyllotaxis, predicated as it is on the view that reductionistic, ontogenetic approaches alone are not adequate. As he says, "La phyllotaxie est un phénomène épigénétique, holiste, systémique; elle opère au-dessus de la chimie et de la physique." This does not, of course, mean that the phenomena of phyllotaxis cannot be analyzed; the question is simply whether the units of analysis postulated by the reductive paradigm are the appropriate ones for approaching phyllotaxis, and morphogenesis in general. The evidence we possess which bears on this basic question indicates that they are not adequate; part of the evidence arises in the systematic critique of phyllotaxis that Dr. Jean provides.

In summary, I believe that this book of Dr. Jean's is an important contribution to an entirely new approach to pattern and form in biological systems. I strongly recommend its careful reading to anyone interested in these questions. *It is a monumental work in a key sector of morphogenesis.*

ROBERT ROSEN
Department of Mathematics
University of Texas at Arlington

PREFACE

*We must study phyllotaxis which is
the bugbear of botany, so simple, yet
so profound as to be incomprehensible.*
(Corner, 1981)

THE SUBJECT

This book deals with mathematical problems raised by the growth of plants, more precisely with the problems related to what is currently called the *PHENOMENON of PHYLLOTAXIS*. This phenomenon, described in the Introduction, is one of the most striking and puzzling characteristics of apical activity. In the past decade many mathematical models have been proposed of the theories dealing with the problem of the origin of the patterns displayed on the plant apex. This book, resulting from a positive critical analysis, brings these models together in order to point out their respective successes, to compare them, and to suggest directions for research toward a better solution of the problem of phyllotaxis, that is, toward a better integration of its many facets. It is a treatise that makes accessible all the recent mathematical developments, as well as those that have appeared in specialized periodicals since 1830. Let us call attention to the masterly works of Bravais and Bravais (1837), van Iterson (1907), Richards (1951), and Adler (1974) in particular.

An increasing number of workers recognize the strategic position of the field within plant morphogenesis. Some consider the problem of phyllotaxis as a test case in biomathematics. For the biomathematician Robert Rosen, "Understanding phyllotaxis may well provide the key for unlocking the ancient mysteries surrounding the generation of pattern and form in the organic world" (from the Foreword). For Dormer (1965), "The study of stem apices is one of the main branches of morphogenetic research." The growing plant apex, one of the most complex of entities, is defined by the botanist Wardlaw as a "geometric dynamical system possessing a biological organization." It thus interests both botanists and mathematicians. It is at the apex that the phenomenon of phyllotaxis is observed, in all its simplicity, as the appearance of consecutive terms of a particular sequence of integers, the so-called Hauptreihe or Fibonacci series $\langle F(k) \rangle$, defined by $F(1) = F(2) = 1$, $F(k + 1) = F(k) + F(k - 1)$, $k > 1$. The question "Why does that sequence arise in the secondary spirals seen on plants?" constitutes the *PROBLEM of PHYLLOTAXIS* and suggests a deep morphogenetic control mechanism, phyletically and ontogenetically.

As pointed out by Alder (in the preface of Jean, 1978b), physics also has its famous series. The determination of the difference in energy between two electronic levels of the hydrogen atom involves the two-parameter sequence $\langle 1/n^2 - 1/m^2 \rangle$. The question "Why does this sequence arise in the spectrum of atoms?" is at the heart of the problems of atomic structure. These were solved as soon as an answer was found to the question, first partially by Bohr's atomic theory, then more completely by quantum mechanics and electrodynamics. The question that is formulated above in plant biology thus appears to be a most stimulating point of departure to study *MORPHOGENESIS* for all the biologists and mathematicians who, without being reductionists, envy the successes of axiomatic mathematical physics.

The subject of the book is central to those studying plant development, plant macroevolution, and aspects of plant breeding and genetics. Biomathematics is now taking its place in university programs. With this increasing mathematization of the life sciences, this book might be used in an advanced course or as a personal teaching and reference tool by those working in related areas, or looking for an inexhaustible domain of applications of mathematics, or wishing to move rapidly into the state of the art of one of the most attractive facets of plant growth.

THE CONTENT

The phenomenon of phyllotaxis is an expression of a fundamental law of organization in plants. Research in this area has taken three different forms: (1) experiment and observation; (2) mathematical modeling that is only descriptive; and (3) mathematical modeling that aims to be explanatory. The *EXPERIMENTAL and OBSERVATIONAL WORK* carried out for more than a century, based on anatomical and physiological approaches, is described fully by Cutter (1965), Loiseau (1969), and Wardlaw (1965a). The main results of these researches, needed for a sound mathematical treatment and an adequate perception of the problem in its entirety, are scattered throughout the text, mainly in Chapter 5 and in the Research Activities at the end of Chapters 2, 3, and 5. These results and the theories molded from them became the foundations of the mathematical models of the 1970s. The References will guide the reader toward a thorough knowledge of the biological aspects of the problems related to phyllotaxis. This book should lead the empiricist to strengthen the data, so that it becomes possible to distinguish between the theories. "It is a chastening thought that, after so many decades of work, we still know so little of the factors which determine phyllotaxis and related phenomena" (Wardlaw, 1968a).

The descriptive problem of phyllotaxis consists of trying to identify parameters and to formulate concepts by which the phenomenon can be reported with precision. In the middle of the nineteenth century the first coherent descriptive mathematical treatment of the phenomenon of phyllotaxis was launched by Bravais and Bravais (1837). The approach of these pioneers, known as the cylindrical representation, was developed by van Iterson (1907) at the beginning of the twentieth century and elaborated by Adler (1974). Richards (1951) introduced an original mathematical description, in the so-called centric representation of the phenomenon, thoroughly analyzed by Jean (1979b). The *DESCRIPTIVE MODELS* are at least the frames of reference in which the third category of research takes place. As such, they are fundamentally important. Before the implications of an explanatory model are explored, it is indeed natural and necessary to try to discover the general properties of the framework in which the model is formulated. Thus the properties of the cylindrical lattice are scrutinized in Chapters 1 and 2, those of the centric representation of phyllotaxis in Chapter 3, and those of the diffusion equations in phyllotaxis in Chapter 4.

The study of phyllotaxis reaches the heart of the problems related to differential growth and allometry in plants. Section 3.3.5 introduces recently uncovered allometric relations regarding the radial spacing of the primordia of plants (Jean, 1983g).

Of course our main objective is centered on the *EXPLANATORY MODELS*, such as those of Adler (1977a), Coxeter (1972), Jean (1980c), Veen (1973), Thornley (1975b), and Young (1978). They are gathered together in Chapter 5. They are based on such theories as Hofmeister's hypothesis (1868), Snow's first available space theory (Snow and Snow, 1962), Schwendener's contact pressure theory (1878), Plantefol's foliar helices theory (1947), Church's phylogenetic theory (1904), Schoute's field theory (1913), and on the vascular level, Bolle's theory of bifurcated induction lines (1939). The Epilogue presents a comparative analysis of these models and theories.

The Research Activities sections are meant to introduce particular publications and to point out specific facts or trends not completely covered by the main text. The Appendixes contain the complete solutions of the Exercises in Chapters 1 and 2. The mathematics used ranges from the Chinese remainder theorem and the properties of continued fractions to partial differential equations and computer simulation. The insistence is not so much on the techniques of problem solving as on the explanation of mathematical concepts proved to be relevant in plant biology.

This is the first complete treatise on phyllotaxis. The book entitled *Phytomathématique* (Jean, 1978b) has a different orientation: it has an intuitive approach and, although it obviously touches on the fascinating problem of phyllotaxis, only mentions the subjects that are presented independently and systematically here. The domain of phyllotaxis has thus become a deductive and articulated system of methods, concepts, and results proper to the permanence of a science now more accessible and open to exploration and development, which I suggest calling *BOTANOMETRY*.

Besides its completely new Preface, the Introduction, the section on allometry which brought a rearrangement of the headings of Chapter 3, a few more Research Activities and comments, a more complete Table of Contents, and the Indexes, the book differs slightly in its text from the French version. Moreover, Appendixes 2 and 3 have been deleted and Appendixes 1 and 4 incorporated into the Introduction and into a Research Activity of Chapter 5.

From the French handwritten manuscript to the present book, the proofs of half a dozen intermediate texts have had to be corrected. Though this processing has brought improvements in the text, and though all possible care has been taken so as to obtain an errorfree book, it would be too optimistic to expect that all errors and inconsistencies (for which I take full responsibility) have been eliminated. I would appreciate it if these, together with any other suggestions for improvement, are brought to my attention.

ROGER V. JEAN

Montreal
January 1984

ACKNOWLEDGMENTS

I am obliged to Professor Rosen, presently professor at the department of physiology and biophysics at the Dalhousie University, for his involvement in the form of a foreword.

Very special thanks go to the courteous and efficient personnel of Wiley-Interscience with whom I have been working, in particular to Mary M. Conway, the Life Sciences Editor. The professionalism of the publisher made the realization of this adapted English version a stimulating experience.

I want to express my gratitude to the Université du Québec à Rimouski for reducing my teaching activities in the winter of 1983, which made possible the preparation of this edition.

I am thankful to Mrs. Jocelyne Desgagnés for her diligent and brilliant typing of the final manuscript.

Finally, I am grateful to The Canada Council for fellowships at the beginning of my research project, to the Natural Sciences and Engineering Research Council of Canada for their continuous support, especially for two three-year grants, and to the Ministère des Affaires Intergouvernementales du Québec for missions of research.

R.V.J.

CONTENTS

*The domain of phyllotaxis
has very recently accomplished
considerable progress through
mathematical modeling*

MATHEMATICAL APPROACH TO PATTERN AND FORM IN PLANT GROWTH

Introduction
THE PHENOMENON
OF PHYLLOTAXIS

It is an astonishing fact — to the writer at least — that, in spiral phyllotaxis, the numbers of contact parastichies are usually two adjacent members of the Fibonacci series.
(Wardlaw, 1968a)

The plant kingdom and the world of mathematics are generally perceived as having very few things in common. The luxuriance of vegetation, the infinite variety of forms, and the diversity of patterns, all challenging the imagination, do not appear to be controlled by mathematical equations. And yet, behind this apparent confusion, mathematical constants hide. Great minds such as Pythagoras, Goethe, and Leonardo da Vinci were aware of that. But it was not until the middle of the nineteenth century that botanists blew wind into the sails. Around 1950 the debate was definitely engaged and publications on the subject began to abound; today their number is growing exponentially. More and more mathematicians devote themselves to biomathematics. *PHYTOMATHEMATICS* is the mathematical approach to plant growth, form, and pattern.

The sunflower, a perfect example of phyllotaxis, at the height of its splendor in July and August, probably calls forth contemplation other than mathematical. However, its poetry is not unrelated to the spiral pattern displayed by the florets of the capitulum of the plant. This spirality is more conspicuous with the coming of fall, when petals die and stamens fall, exposing the fruits distributed on two families of spirals winding in opposite directions. This is an interesting fact in itself, but more surprising is the number of spirals in each family. Certain numbers crop up, excluding others, almost without exception, that is, 13, 21, 34, 55, 89, and 144. For example, the diagram on the right in Figure 1 is a representation of the flower head on the left, displaying the pattern denoted by 34/21. Church observed in 1899 at Oxford a huge sunflower head, 56 cm in diameter, with 144 + 233 parastichies; O'Connell (1951) reports he has found this pattern on a specimen in Vermont.

The ancients noticed the spiral architecture of the buds of daisies and sunflowers. Botanists usually call these spirals parastichies. One also finds them on pinecones, where the numbers are 2, 3, 5, 8, and 13 (Brousseau, 1968). On the surface of the pineapple there are generally three families of such spirals, because of the hexagonal shape of the scales. The observer may count them, as a worker in Hawaii reported in the *Pineapple Quarterly*. The numbers he obtained, over a period of two years, without exception, are 5, 8, 13, 21, and 34, depending on the size of the fruit. Such observations have been systematically made by Davis (1970, 1971) on the trunks of palm trees, where the palm leaf traces determine the parastichies. More generally, looking at a plant apex or a transverse section of it, one ordinarily sees the leaf *PRIMORDIA* (the future leaves or scales) organized into two families

2

of spirals, the so-called contact or conspicuous parastichies—one family going up to the left, the other going up to the right. The phenomenon of phyllotaxis is characterized (1) by the fact that the numbers n and m of parastichies in the two families are, in almost all cases (95%), consecutive terms of the *MAIN SERIES* 1, 1, 2, 3, 5, 8, 13, 21,..., and (2) by the convergence of the divergence angle, that is, the angle between consecutively emerging primordia, to the angle ϕ^{-2}, that is, 137.51°. The number ϕ (after Phidias, the great Greek Sculptor of the fifth century BC) has the value $(\sqrt{5} + 1)/2 \simeq 1.618$. Why is ϕ^{-2} generally encountered? Why are the numbers 4, 6, 7, 9, 10, 11, 12, 14, 15, 16, 17, 18, 19, 20, and 22 so rarely found? These remarkable facts call for an explanation.

Many facets of the organization of plants depend on the determination of the position of their primordia, and *PHYLLOTAXIS* is the study of their relative arrangements. More precisely, it is the study of plant growth in the sense of addition to an organism of new parts, similar to those already existing, such as the scales of cones, the florets of the capitulum of Compositae, the branches, leaves, and buds of trees, the primordia of seedlings, the foliar traces of the vascular system of plants, the sporophylls (pistils, stamens: anthotaxis), and the semaphylls (petals, sepals, bracts: semataxis). More generally, phyllotaxis is the study of the symmetry

Figure 1   Generating a 34/21 capitulum of a chrysanthemum by means of two families of logarithmic spirals, irradiating around a common pole, 34 in one direction, 21 in the other. (Diagram on right from *The Divine Proportion, by* H. E. Huntley, 1970, Dover Publications, Inc., N.Y.).

properties of plants in relation to their functions in their natural environment (reproduction, photosynthesis, phylogenesis). Finally it is the study of relative growth rates, allometry, and morphogenesis in the apical meristem.

The following experiment illustrates one of the first observations made by investigators in phyllotaxis. On a leafy stem, the leaves appear to be arranged at random. Let us attach a thread to the petiole of a leaf, near the stem, and wind it up around the stem by the shortest path, from one leaf to the next, until we reach a leaf that appears approximately superposed on the first, such as leaf 3 and leaf 11 in Figure 2. In the majority of cases the ratio of the number of turns around the stem to the number of leaves met, excluding the first, is one of the fractions of the following sequence made with the Fibonacci numbers:

$$1/2, 1/3, 2/5, 3/8, 5/13, \ldots,$$

For example, in Figure 2 the foliar cycle is defined by the fraction $3/8$, as

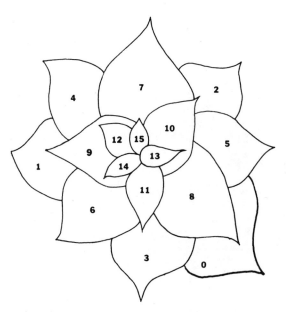

Figure 2 An overview of a leafy stem. To follow the shortest path through the intermediate leaves between two leaves approximately superposed, such as leaves 7 and 15, or 3 and 11, we must make three turns around the stem and we meet eight leaves: $3/8$ is the phyllotactic fraction of the system.

for the poplar and the pear tree. For the apple tree and the cherry tree the fraction is $2/5$.

This discovery by Schimper (1836)—the word phyllotaxis comes from him—fostered an enthusiasm that was translated into many publications, such as those of Bravais and Bravais (1837), but also, unfortunately, an idle story that strongly influenced morphogenetic ideas for almost 100 years. The preceding fractional notation is just a descriptive terminology meant to represent the angle between consecutive leaves in the bud, where precise measurements can be taken. These fractions are thus approximations of initial relations, before the deformations following the growth of the stem. But the preceding series of fractions of a turn between consecutive leaves rapidly converges to ϕ^{-2}. It follows that the numbers $m = F(k + 1)$ and $n = F(k)$ of opposed parastichies are related to the divergence angle ϕ^{-2} by the equation $\phi^{-2} = 1 - 1/\lim_k[F(k + 1)/F(k)]$.

Even the exceptional cases are governed by the same recurrence relation $u_{n+1} = u_n + u_{n-1}$ that characterizes the Fibonacci sequence, except that it is applied to different initial terms. An example is Schoute's accessory series $1, 3, 4, 7, 11, 18, 29, \ldots$, called Lucas' sequence by the mathematicians, for which the divergence angle is $(3 + \phi^{-1})^{-1}$. For the normal phyllotaxis sequence $1, t, t + 1, 2t + 1, 3t + 2, \ldots$, the divergence angle is $(t + \phi^{-1})^{-1}$. Anomalous cases are known, although rare, where the sequence is $2, 5, 7, 12, 19, 31, \ldots$.

What seems anarchic is in fact well organized. The problem of phyllotaxis is to identify the regulatory process that restricts the numbers m and n to those given. This regulation must first be translated into mathematical terms, and then reproduced in explanatory models: this is the program of the book.

1 THE MATHEMATICAL FOUNDATIONS

In every theoretical field, whether it be physics, biology or economics, particular systems of concepts and models appear
(Fomin and Berkinblitt, 1975)

*This chapter is used with permission from Consortium for Mathematics and Its Applications, Inc., 271 Lincoln Street, Suite Number 4, Lexington, Massachusetts 02173. It is taken from the author's UMAP module U571, "The Use of Continued Fractions in Botany," (1981b).

7

1.1 PRESENTATION

This chapter presents a few of the mathematical notions derived from the observation of plants that are elements of the language used in botanometry. First a formula to approximate the divergence angle between consecutive primordia of a plant is developed, and, consequently, the mathematical foundations of the phenomenon of phyllotaxis are established.

The approximation formula can be found by scrutinizing an article by Bravais and Bravais, published in 1837 in a French periodical. This text, which marks the birth of botanometry, had been almost forgotten when Tait resuscitated it in a very short article published in 1872. Recently Jean (1981b)* took this formula out of the context of periodicals meant for narrow circles of specialists and made a rigorous and pedagogically oriented presentation of it (Section 1.3.1). It makes readily available an important result (Section 1.3.2) concerning the relation between two major notions: the divergence and the phyllotaxis of a plant (Sections 1.2.1 and 1.2.2). The complete relation is produced in Chapter 2, in the context of visible opposed parastichy pairs.

Section 1.4 offers a first explanation of the phenomenon of phyllotaxis, that is, of the emergence in nature of the angles $(t + \phi^{-1})^{-1}$ together with the series $\langle S_{t,k} \rangle$, a particular case of which is the almost omnipresent divergence angle $1/\phi^2$ and the main series $\langle F(k) \rangle$, defined earlier. The key to this presentation is Klein's theorem (Section 1.2.4) on continued fractions (Section 1.2.3). This theorem enables us to reach rapidly the essence of Coxeter's contribution (1969, 1972) to botanometry, regarding the role of the intermediate convergents of continued fractions.

For an adequate understanding it may be desirable to go through the exercises at the end of each section before going on to the next section; these exercises quite often contain central results.

1.2 FUNDAMENTAL CONCEPTS

1.2.1 Phyllotaxis of a System

We have seen that the pineapple, the sunflower, the pinecone, and the daisy, for example, show two families of *PARASTICHIES* winding in opposite directions. Figure 1.1 is a transverse section of a bud of a pine, showing the

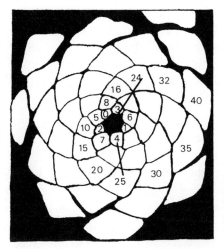

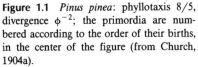

Figure 1.1   *Pinus pinea*: phyllotaxis 8/5, divergence ϕ^{-2}; the primordia are numbered according to the order of their births, in the center of the figure (from Church, 1904a).

so-called primordia, here the future scales, distributed on eight clockwise and five counterclockwise parastichies. Table 1.1 gives the numbers of parastichies on sunflowers, according to their sizes. These observations lead to an important concept, called the *PHYLLOTAXIS* of a plant. It is a pair of integers, denoted by m/n, $m > n$, where m and n are the numbers of conspicuous opposed parastichies determined by the *PRIMORDIA* (scales, florets, leaves, etc.) of the plant. A plant always shows many pairs of opposed parastichies: the phyllotaxis is defined by the conspicuous pair,

Table 1.1   The phyllotaxis of sunflowers (from Jean, 1978b)

Size of the head of the sunflower	Number of parastichies	
	Positive direction	Negative direction
Very small	13	21
Small	21	34
Normal (14–15 cm in diameter)	34	55
Large	55	89
Very large	89	144
Huge ($\approx$ 55 cm in diameter)	144	233

also called, in an expressive way, the *CONTACT PARASTICHIES*, derived from the contacts between the primordia. Everywhere in the text, except in Figures 1.1 and 2.2 where 0 is the youngest, the oldest primordium is denoted by 0. The former notation, still in use by some authors, has the disadvantage of calling for a new numbering of the primordia each time a primordium arises.

1.2.2   Divergence and Phyllotactic Fraction

Another parameter that comes out of the observation of plants, and that is important for their description, is the divergence angle, or simply the divergence. Most of the time this angle is equal to 137°30′27″. The *DIVERGENCE* of a plant is the angle at the center, in the transverse section of the bud, determined by the centers of two consecutively born primordia. It is the angle between the points 24 and 25 in Figure 1.1. Most

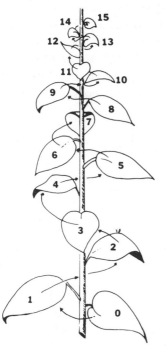

Figure 1.2   An illustration of the notion of phyllotactic fraction. Corresponds to Figure 2 of the Introduction.

of the time this angle is expressed as a part of a turn, that is, as a number between 0 and 1, more precisely between 0 and $\frac{1}{2}$.

The divergence observed in a bud, at the origin of the growth, appears to be an irrational number, the primordia being distributed on spirals. If the bud becomes a leafy stem, it is possible to obtain an approximation to the value of divergence by marking two leaves appearing approximately superposed. They determine, such as the leaves 3 and 11 in Figure 1.2, what is called an *ORTHOSTICHY*, a line approximately parallel to the axis of the cylindrical stem. In a movement up around the stem, following the *GENETIC SPIRAL** determined by the shortest path through two consecutively born primordia or leaves, one goes from leaf 3 to leaf 11 by three turns around the stem, and eight leaves. The value $3/8$ is an approximation to the divergence. This brings us to the following definition. A *PHYLLOTACTIC FRACTION* of a leafy stem (or of a plant in a cylindrical lattice; see Sections 1.3.2 and 1.4.1), obtained by determining two leaves on an orthostichy, is the ratio of the number of turns around the stem to the number of leaves met along the portion of the genetic spiral between the two leaves (excluding the first). Experiments show that the most common phyllotactic fractions are the members of the sequence shown in Exercises 1.4 and 1.5.

1.2.3 Convergents of a Continued Fraction

Every positive real number ω can be written, using Euclid's algorithm, in the form

$$\omega = a_0 + \cfrac{1}{a_1 + \cfrac{1}{a_2 + \cfrac{1}{a_3 + \cdots}}} \tag{1.1}$$

where $a_0, a_1, a_2, \ldots$ are positive integers. To obtain this development we separate out the largest positive integer a_0 contained in ω and write

$$\omega = a_0 + r_0, \qquad \text{where } 0 \leq \bar{r}_0 < 1.$$

*Also called the fundamental, generative or ontogenetic spiral.

If $r_0 \neq 0$, then we treat $1/r_0$ as we did ω:

$$\frac{1}{r_0} = a_1 + r_1, \quad \text{where } 0 \leq r_1 < 1,$$

and continue in the same way:

$$\frac{1}{r_1} = a_2 + r_2, \quad \text{where } 0 \leq r_2 < 1,$$

$$\frac{1}{r_2} = a_3 + r_3, \quad \text{where } 0 \leq r_3 < 1,\ldots .$$

According to whether ω is rational or not, the process terminates or goes on indefinitely. An infinite continued fraction is written $\omega = [a_0; a_1, a_2, \ldots]$, and a terminating continued fraction is written $\omega = [a_0; a_1, a_2, \ldots, a_m]$.

The *CONVERGENTS* or *PRINCIPAL CONVERGENTS* of ω are the rational numbers $[a_0; a_1, a_2, \ldots, a_k]$, $k = 0, 1, 2, 3, \ldots$, denoted by p_k/q_k. In the case where ω is rational, $0 \leq k \leq m$. Obviously we have

$$p_0 = a_0, q_0 = 1, p_1 = a_0 a_1 + 1, q_1 = a_1. \tag{1.2}$$

The following results give inductively the values of p_k and q_k, for every $k \geq 2$.

Laws of formation of the convergents:

$$p_k = a_k p_{k-1} + p_{k-2},$$
$$q_k = a_k q_{k-1} + q_{k-2}. \tag{1.3}$$

It can be proved that the convergents of even order p_{2n}/q_{2n}, $n = 0, 1, 2, \ldots$, form an increasing series converging toward ω, and that the convergents of odd order form a decreasing series also converging toward ω. The *INTERMEDIATE CONVERGENTS* of ω, when they exist, are the ratios $p_{k,c}/q_{k,c}$, where $0 < c < a_k$, $k \geq 2$, and c is an integer. These convergents satisfy the equations

$$p_{k,c} = cp_{k-1} + p_{k-2},$$
$$q_{k,c} = cq_{k-1} + q_{k-2}, \quad k > 1. \tag{1.4}$$

1.2.4 Klein's Diagram

In 1896 Klein gave a remarkable geometric interpretation of the way the convergents of the continued fraction of an irrational number converge toward that number. Let us imagine, with him, pegs or needles inserted at all points of the (x, y) plane having integral and nonnegative coordinates. We obtain a *SQUARE LATTICE* of well-aligned pegs, like the trees in an orchard. Looking from the origin of coordinates, we see points of the lattice in all rational directions, and only in such directions. The field of view is everywhere "densely," but not completely and continuously, filled with "stars." We might be inclined to compare this view with that of the Milky Way (Klein, 1932). The number-theoretic properties of the convergents deliver the following result, illustrated in Figure 1.3.

Klein's Theorem. Given the square lattice and the irrational number $\omega > 0$, wrap tightly drawn strings about the sets of pegs to the right and to

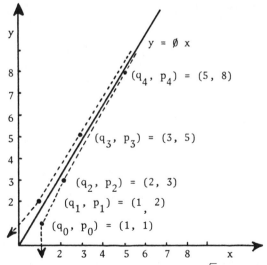

Figure 1.3 Klein's diagram for the continued fraction of $\phi = (\sqrt{5} + 1)/2$ in the square lattice, showing the first few vertices of the two infinite polygonal lines (see Exercise 1.8).

the left, respectively, of $y = \omega x$; then the vertices of the two infinite string-polygons that bound the two sets are the points (q_i, p_i) whose coordinates are respectively the denominators and the numerators of the convergents p_i/q_i of the continued fraction of ω, the right infinite polygon having the even convergents, and the other the odd convergents.

Exercises

1.1 Obtain specimens of cones and pineapples and determine their respective phyllotaxis m/n. The types of phyllotaxis occurring here belong to the *MAIN PHYLLOTACTIC SERIES* $2/1, 3/2, 5/3, 8/5, 13/8, \ldots$.

1.2 Prove that the fraction of a turn $1/\phi^2$, where $\phi = (\sqrt{5} + 1)/2$, corresponds to the divergence angle mentioned in Section 1.2.2.

1.3 Show that the divergence $1/\phi^2$ is related to the phyllotaxis $F(k + 1)/F(k)$ (of the main phyllotactic series) by the formula

$$\frac{1}{\phi^2} = \lim_{k \to \infty} \left[1 - \frac{1}{F(k+1)/F(k)} \right]. \tag{1.5}$$

1.4 Determine the phyllotactic fractions of the leafy stems of your favorite trees.

1.5 Show that *SCHIMPER AND BRAUN'S SERIES* $1/2, 1/3, 2/5, 3/8, 5/13, \ldots$ converges toward $1/\phi^2$.

1.6 Show that $\phi = 1 + \cfrac{1}{1 + \cfrac{1}{1 + \cfrac{1}{1 + \cdots}}}$,

$$\sqrt{2} = 1 + \cfrac{1}{2 + \cfrac{1}{2 + \cfrac{1}{2 + \cfrac{1}{2 + \cdots}}}}.$$

1.7 Show that, for all $k \geq 0$,

$$q_k p_{k-1} - p_k q_{k-1} = (-1)^k. \tag{1.6}$$

1.8   Show that the successive convergents of ϕ and of $1/\phi^2$ are, respectively, the terms of the main phyllotactic series and of Schimper and Braun's series. Show that the convergents of $(t + \phi^{-1})^{-1}$ are $F(k)/S_{t,k}$, where

$$S_{t,k} = F(k)t + F(k-1), \qquad (1.7)$$

and t is a positive integer, $F(0) = 0$, $F(-1) = 1$, $F(k+1) = F(k) + F(k-1)$.

1.9   Show that every real number ω, with no intermediate convergent, has the form $\omega = [a_0; a_1, 1, 1, 1, 1, \ldots]$.

1.10   Write the first few convergents of $\sqrt{5}$ and of $1/(e-1)$. Verify that $67/29 = [2; 3, 4, 2]$.

1.11   Construct Klein's diagram for the continued fraction of $1/(e-1)$. Find two points on the infinite polygonal lines that are not vertices.

1.12   Let $\omega = [a_0; a_1, a_2, \ldots]$ and p_n/q_n and $p_{n,c}/q_{n,c}$ be its convergents. Show that the points $(q_{n,c}, p_{n,c})$ of the square lattice are on the segment of the infinite polygonal line determined by the points (q_{n-2}, p_{n-2}) and (q_n, p_n).

1.3   APPROXIMATION FORMULA FOR THE DIVERGENCE

1.3.1   Illustration of Bravais' Formula

This formula gives a good idea of the value of the divergence d in a system whose phyllotaxis is known:

$$d \simeq \frac{cp + sq}{cm + sn}. \qquad (1.8)$$

The ratio in Expression 1.8 is a phyllotactic fraction determined by six parameters defined in the following way:

1. m/n is the phyllotaxis of the system.
2. p/q is the convergent of the continued fraction of m/n satisfying one of the relations $mq - np = \pm 1$ for which $d < \frac{1}{2}$; $p < m$ and $q < n$ can also be determined by an examination of the cylindrical lattice corresponding to the plant under consideration (Section 1.3.2).

3. c and s are the numbers satisfying the relation $cm + sn = k$, where k is the number of primordia on the genetic spiral between those two determining the chosen orthostichy; these parameters can also be determined experimentally or by an examination of the cylindrical lattice, as the following example shows.

Before proving the formula, in the next section, let us take a cone like the one illustrated in Figure 1.4 and determine its phyllotaxis m/n, $m > n$. Here we have $m/n = 8/5$, that is, a family of eight counterclockwise parastichies and a family of five clockwise. Take a scale near the base of the cone and put the number 0 on it. Inscribe the numbers 5 and 8 on the two neighboring scales belonging to the opposed contact parastichies meeting at scale 0. The figure shows how the other scales must be numbered, according to the following theorem, corresponding to the order of their births.

Bravais' Theorem (Bravais and Bravais, 1837). The numbers on the consecutive scales of any given parastichy differ by the number of parastichies in the family of the given one.
Figure 1.4 also shows two *ORTHOSTICHIES*. These are straight lines, approximately parallel to the axis of the cone, determined by nearly superposed scales. The scales $0, 21, 42, 63, \ldots$ form an orthostichy, and the scales $0, 34, 68, 102, \ldots$ another.

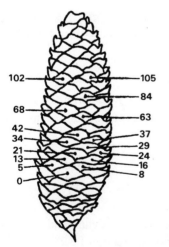

Figure 1.4 The numbering of the scales of a cone according to Bravais' theorem.

Given a cone and the numbering of its scales, let us now determine the values of the six parameters found in Expression 1.8. For definiteness we take the orthostichy going through the scales $0, 21, 42, 63, \ldots$, for which $cm + sn = 21$, that is, $8c + 5s = 21$. This is a diophantine equation whose only solution is $c = 2$ and $s = 1$. Notice that the examination of the cone gives these values, since scale 21 is on the second parastichy parallel to the one passing through the scales 0 and 5, and on the first parastichy parallel to the one passing through the scales 0 and 8. The convergents of the continued fraction of 8/5 are 1/1, 2/1, 3/2, and 8/5. The convergent 3/2 gives $p = 3 < m$, $q = 2 < n$, $mq - np = 1$, and $(cp + sq)/(cm + sn) < 1/2$. The approximation for d is thus 8/21 (see Exercise 1.13).

1.3.2 Jean's Phyllotaxis-Divergence Relation

Figure 1.5 represents the cone of Figure 1.4 considered as a cylinder sectioned along the generator going through the scale 0 and unrolled in the plane. Scale 0 thus appears at a and at A. The spirals linking the scales 0 and 5 and the scales 0 and 8 become straight lines meeting at scale 40. They determine the triangle aBA, called the *VISIBLE OPPOSED PARASTICHY TRIANGLE* belonging to the pair (5, 8) (Adler, 1974). There are five lines parallel to $\overline{AB}$ and meeting all the scales, and eight lines parallel to $\overline{aB}$ and doing the same. For simplicity and without loss of generality, it is supposed that $\overline{aA} = 1$. The scales are thus represented by their centers in an infinite vertical strip of well-aligned points. This is the so-called *NORMALIZED CYLINDRICAL LATTICE*, where the *DIVERGENCE* of a plant is the difference between the abscissas of two consecutive scales on the genetic spiral, and the ordinate of point k is kr, with $r > 0$. Notice that the coordinates of scale 1 are (d, r). We show, in Section 1.4.1, the relation between this lattice and the square lattice of Section 1.2.4.

The following argument, which for definiteness is based on the example, produces Bravais' formula. Segment $\overline{AE}$ in Figure 1.5, approximately vertical,

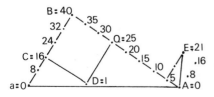

Figure 1.5 The visible opposed parastichy triangle for the pair (5, 8) (from Jean, 1978b).

represents the orthostichy linking the scales 0 and 21. The number of complete revolutions around the axis of the cone, to go from scale 0 to scale 21, is equal to $\{(cm + sn)d\}$, where c and s are such that $cm + sn = 21$, and $\{\cdot\}$ means the integer nearest to $(cm + sn)d$. It follows that the phyllotactic fraction $\{(cm + sn)d\}/(cm + sn)$ approximates d. The approximation is improved when the value of the expression

$$|(cm + sn)d - \{(cm + sn)d\}|$$

is smaller, that is, when the orthostichy is closer to the vertical axis through A. By the properties of the (approximately) similar triangles aCD and aBA we have

$$d \cong \frac{\overline{aC}}{nl} \cong \frac{\overline{CD}}{ml^*} \tag{1.9}$$

(respectively equal to $2/5$ and $3/8$ in the example), where l and l^* are the lengths of the regular *STEPS* on the opposed parastichies, that is, the distances between scales 0 and 8, and between scales 0 and 5, respectively. Putting $\overline{aC} = ql$ and $\overline{CD} = pl^*$, we have

$$\{(cm + sn)d\} \cong \{cp + sq\} = cp + sq. \tag{1.10}$$

It follows that $d \cong (cp + sq)/(cm + sn)$. The same approximation is obtained when comparing triangles DQA and aBA (note that the meaning of p and q changes however). It remains to be shown that p/q is a convergent of the continued fraction of m/n. Since $mq - np = \pm 1$ (see Exercises 1.16) the result follows (compare to Exercise 1.7).

Determination of p, q, c, and s from the cylindrical lattice. These parameters can be obtained from the inspection of the visible opposed parastichy triangle. The parameter q is the number of steps in the direction of the m-parastichies, to go from a to C, or from D to Q, depending on whether one compares triangle aBA to triangle aCD or to triangle AQD, in order to obtain $d < \frac{1}{2}$, that is, depending on the sense of the genetic spiral. The parameter p is the number of steps in the direction of the n-parastichies, to go from D to C or from A to Q, according to the previous choice of triangles. The parameters c and s are the number of steps in the direction of the m-parastichies and n-parastichies, respectively, to go from a or A (irrelevant) to the point determining the orthostichy. From the above we can deduce the following theorem (see Exercises 1.17 and 1.18) relating the divergence to the phyllotaxis in a system of primordia.

Theorem (Jean, 1981b). If $m/n = S_{t,k}/S_{t,k-1}$, $t \geq 2$, then d is equal to $F(k)/S_{t,k}$ or to $F(k-1)/S_{t,k-1}$ or is between these two values. When k tends toward infinity, $d = (t + \phi^{-1})^{-1}$.

Exercises

1.13   Give another approximation for d in the cone of Figure 1.4.

1.14   Apply the material in this section to various specimens of cones and pineapples: number the scales, draw the visible opposed parastichy triangle, determine the parameters p, q, c, and s by different methods, draw some orthostichies, apply Bravais' formula, and compare the results.

1.15   Make your own presentation of the approximation formula.

1.16   Referring to the symbolism used in this section prove that

$$mq - np = \pm 1. \tag{1.11}$$

1.17   Referring to Figure 1.5, show that

$$p/m \leq d \leq q/n \quad \text{or} \quad q/n \leq d \leq p/m. \tag{1.12}$$

1.18   Prove the theorem presented in Section 1.3.2.

1.19   Verify that the values for d obtained in Exercise 1.14 satisfy the theorem of the preceding exercise.

1.4   A FIRST EXPLANATION OF PHYLLOTAXIS

1.4.1   Bravais' Cylindrical Lattice in Botany Versus Klein's Square Lattice for Continued Fractions

Consider the following transformation of the (x, y) plane of Section 1.2.4 into the (X, Y) plane of the cylindrical lattice of Section 1.3.2:

$$X = \omega x - y, \quad Y = rx, \quad \text{where } r > 0. \tag{1.13}$$

The line $y = \omega x$ of the (x, y) plane becomes the line $X = 0$, that is, the Y-axis of the (X, Y) plane. The vertices (q_i, p_i) of the infinite polygons of the continued fraction of ω in Klein's square lattice become the vertices $q_i = (\omega q_i - p_i, rq_i)$ of two infinite polygonal lines, asymptotic to the Y-axis and containing between them only one image of that lattice, that is, the point $(0, 0)$. All the points of the lattice on these polygonal lines are called the *NEIGHBORS OF THE Y-AXIS*. In the region $0 \le X < 1$, $Y > 0$, the coordinates of the points corresponding to those of the square lattice are

$$\left(\omega x - y - [\omega x - y], rx \right) = \left(\omega x - [\omega x], rx \right), \tag{1.14}$$

where $[\cdot]$ means the integral part, and where x and y are positive integers. Those two or three points that correspond to those of the square lattice and that are nearest to $(0, 0)$ in the (X, Y) plane are called the *NEIGHBORS OF THE ORIGIN*. The lattice of points with integral coordinates becomes an infinite vertical strip in the region $0 \le X < 1$, which repeats itself in all other regions $n \le X < n + 1$, where n is an integer. The points $(0, 0)$ and $(1, 0)$ of the (X, Y) plane represent the same scale or primordium, point 1 is given by $(\omega - [\omega], r)$, and we have the cylindrical lattice met in Section 1.3.2.

In the light of Klein's theorem of Section 1.2.4 we are now better able to understand by means of the following theorem (Exercise 1.20) why, as a pineapple or a cone grows, the phyllotaxis goes from m/n to $(m + n)/m$, where m and n are consecutive numbers in the main series $\langle F(k) \rangle$, and why a better approximation of the divergence $1/\phi^2$ is obtained by taking orthostichies made with higher numbers in the same series.

Theorem. If $\omega = \phi$, then the point $(F(k), F(k + 1))$ of Klein's diagram for ω in the square lattice becomes the neighbor $F(k) = ((-1)^{k+1}\phi^{-k}, F(k)r)$ of the Y-axis in the cylindrical lattice. As r decreases the neighbors of the origin are consecutive members of the main series and the orthostichy determined by the origin of coordinates and $F(k)$ tends to coincide with the Y-axis as k tends towards infinity.

Figure 1.6 shows the cylindrical lattice corresponding to Klein's diagram for $y = \phi x$. The coordinates of point 1 are (ϕ^{-1}, r), or $(-\phi^{-2}, r)$. The figure shows the rhythmic alternation of the neighbors $F(k)$, $k = 5$–11, of the Y-axis about that axis. As r decreases $F(k + 2)$ replaces $F(k)$ as a neighbor of the origin so that the phyllotaxis of the system is always made of consecutive members of the main series, as is emphasized in the preceding

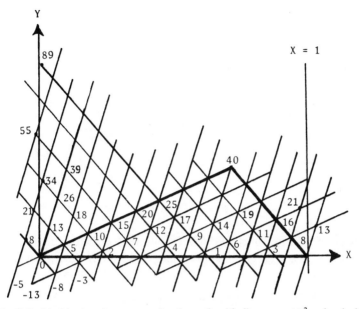

Figure 1.6 Cylindrical lattice of a cone, or of a pineapple with divergence ϕ^{-2} and a clockwise genetic spiral. The points 5, 8, 13 are neighbors of the origin $(0,0)$, or $(1,0)$. The phyllotaxis rises from $8/5$ to $13/8$ with the growth of the fruit, or the decrease of r. The heavy lines are the sides of the visible opposed parastichy triangle belonging to the pair $(8,5)$ (in Figure 1.5 the genetic spiral is counterclockwise) (from Jean, 1978b).

theorem. Notice also, in the figure, that $F(k + 1)$ is closer than $F(k)$ to the Y-axis.

1.4.2 Coxeter's Formula and the Hexagonal Scales of the Pineapple

To begin the derivation of the formula we must make the following observation in the cylindrical lattice of Figure 1.6. The points $F(k - 1)$, $F(k)$, and $F(k + 1)$ are neighbors of the origin if r is such that the angle made by the points $F(k - 1)$, 0, and $F(k)$ is obtuse, and the angle made by the point $F(k)$, 0, and $F(k + 1)$ is acute. This is shown in Figure 1.6 with the points 5, 8, and 13, and in the figure of Appendix 1 with the points 3, 5, and 8. Such a value of r is between two values r_k and r_{k+1} where the two angles are right. One can show (see Exercise 1.24) that

$$r_k = [F(k - 1)F(k)]^{-1/2}\phi^{-k+1/2},$$

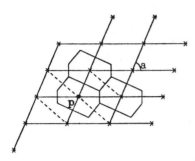

Figure 1.7 Dirichlet's regions corresponding to points of the cylindrical lattice. Here they are hexagonal, as the scales of the pineapple usually are (from Jean, 1978b).

and

$$r_{k+1} = [F(k)F(k+1)]^{-1/2}\phi^{-k-1/2},\qquad(1.15)$$

a formula found in Coxeter (1969). It follows that the three given points are neighbors of the origin if an appropriate value of r is chosen between r_k and r_{k+1}, such as

$$r = \frac{1}{F(k)\phi^k}.$$

Let us look now at the concrete significance of this value of r. We have said that the points of the cylindrical lattice can be considered as the centers of the scales of a pineapple. They determine a tessellation of congruent parallelograms or rectangles called the *FUNDAMENTAL REGIONS*, filling the plane without holes. Each of these points, such as point p in Figure 1.7, belongs to six triangles, as is easily seen by looking carefully at the figure. If the points of intersection of the six perpendicular bisectors of the sides of the triangles meeting at p are joined, what is called the *DIRICHLET'S REGION* of point p is obtained. It is a polygon whose interior is nearer to point p than to any other point of the lattice. These regions are rectangles or hexagons according to whether the angle a is right or not; then point p has four or six neighbors, respectively, and the fundamental region is a rectangle or a parallelogram. It follows that the choice of r between r_k and r_{k+1}, where the scales are rectangles, is such that the scales are hexagons, as they usually are on the pineapple. In particular, we have three families of 8, 13, and 21 conspicuous parastichies for $r = 1/F(7)\phi^7$. Then the coordinates of scale n of the pineapple unrolled in the plane are $(\phi n - [\phi n], n/F(7)\phi^7)$; those of scale 13 are $(1/\phi^7, 1/\phi^7)$.

1.4.3 Normal Phyllotaxis

Figure 1.8 shows a cylindrical lattice where the neighbors of the origin are the points 4, 7, and 11, and the neighbors of the Y-axis belong to the secondary series $1, 3, 4, 7, 11, 18, 29, 47, 76, \ldots$. The phyllotaxis of the system is $7/4$ and may soon become, with the growth of the plant, $11/7$. Obviously this type of phyllotaxis, occurring in nature, is different from the one discussed in the preceding section, arising in 95% of the cases and characterized by the main series $\langle F(k) \rangle$ and the divergence angle ϕ^{-2}. Other types of phyllotaxis, for example those defined by the secondary series $1, 4, 5, 9, 14, 23, \ldots$ and $2, 5, 7, 12, 19, 31, \ldots$ can also be observed, this last case being very rare, however. In almost all cases, the series obtained are of the type $\langle S_{t,k} = F(k)t + F(k-1) \rangle$, $d = (t + \phi^{-1})^{-1}$, $t \geq 2$, and the phyllotaxis $S_{t,k}/S_{t,k-1}$ is said to be normal.

Let $\omega = (t + \phi^{-1})^{-1}$ in the square lattice of Section 1.2.4. Then in the cylindrical lattice, the neighbors of the Y-axis are the points $S_{t,k}$ (see Exercise 1.8). The abscissa of $S_{t,k}$ is $S_{t,k}(t + \phi^{-1})^{-1}$ minus the integer nearest to this expression, that is, $(-1)^k/\phi^{k-1}(\phi t + 1)$ (see Exercise 1.26).

It follows that $S_{t,k}$ and $S_{t,k-1}$ make a right angle with the origin for

$$r_{k+1} = \frac{\left(S_{t,k} S_{t,k-1}\right)^{-1/2} \phi^{-k+3/2}}{\phi t + 1}, \qquad (1.16)$$

a formula that generalizes Coxeter's formula ($t = 1$). The same calculations

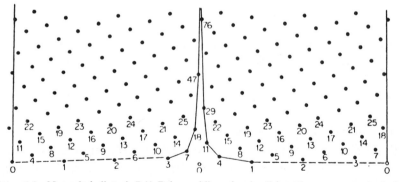

Figure 1.8 Normal phyllotaxis $7/4$. Polygonal lines showing Schoute's accessory series 1, 3, 4, 7, 11, 18,... defined by the same recurrence relation as the main series, from different initial terms (from Coxeter, 1972).

performed with the neighbors $S_{t,k-1}$ and $S_{t,k-2}$ yield r_k. Finally we choose, as we did in Section 1.4.2,

$$r = \frac{1}{S_{t,k-1}\phi^{k-2}(\phi t + 1)},$$

a relevant value for which $S_{t,k}$, $S_{t,k-1}$, and $S_{t,k-2}$ are neighbors of the origin. In the case of Figure 1.8 we have $t = 3$, $k = 4$ and we can calculate that $r \simeq 1/100$, that is, the value used by Coxeter to build the lattice.

Why do the divergences $(t + \phi^{-1})^{-1}$ arise almost exclusively? A partial explanation can be based on observations made by the French botanists Bravais and de Candolle in the nineteenth century, and rediscovered by Coxeter (1972). These observations can be summarized in the following way: the divergences $d < \frac{1}{2}$ encountered in nature are irrational numbers, and the neighbors of the Y-axis in the cylindrical lattice alternate on either side of this axis, such as $8, 13, 21, 34, 55, 89, \ldots$ in Figure 1.6. It follows that d cannot have intermediate convergents. Indeed, if $a_n > 1$, for $n \geq 2$ in the continued fraction of d, then by Exercise 1.23, going upward along the Y-axis, if q_{n-1} is on the left,

$$q_{n,1}, q_{n,2}, \ldots, q_{n,a_n-1}$$

are also on the left (these points are on the segment determined by q_{n-1} and q_{n+1}, according to Exercise 1.12), and q_n appears on the right, followed by

$$q_{n+1,1}, q_{n+1,2}, \ldots, q_{n+1,a_{n+1}-1},$$

and finally q_{n+1} appears on the left of the Y-axis. This contradicts one of the observations. By Exercise 1.9 we have $d = [0; t, 1, 1, 1, \ldots]$.

Exercises

1.20 Prove the theorem of Section 1.4.1.

1.21 With $\omega = \phi$, as in the theorem of Section 1.4.1, calculate the interval of values in which r must be chosen so that in the normalized cylindrical lattice 5 and 3 are neighbors of the origin. Draw the lattice. For which value of r does point 8 replace point 3 as a neighbor of the origin, that is, does the phyllotaxis pass from 5/3 to

8/5? Calculate the value of r for which $F(k + 1)$ replaces $F(k - 1)$ as a neighbor of the origin.

1.22 Show that if the Y-axis contains only one point of the cylindrical lattice, that is, $(0, 0)$, then the divergence is irrational.

1.23 Show that the neighbors of the Y-axis in the cylindrical lattice satisfy the relations

$$q_{n-1} < q_{n,1} < q_{n,2} < \cdots < q_{n,a_n-1} < q_n < q_{n+1,1} < \cdots < q_{n+1}$$

(these numbers are the denominators of the convergents, principal and intermediate, of the continued fraction of ω).

1.24 Prove Coxeter's formula.

1.25 In the lattice of Figure 1.8, show that $d = 1/\phi\sqrt{5} = 1/(3 + \phi^{-1})$.

1.26 Referring to Section 1.4.3, show that the abscissa of $S_{t,k}$ is $(-1)^k/\phi^{k-1}(\phi t + 1)$. Verify that $t = 1$ gives the abscissa of $F(k)$, as in the theorem of Section 1.4.1.

1.27 In the preceding exercise, show that the distance of $S_{t,k}$ to the Y-axis is, up to a constant factor, the distance of $(S_{t,k}, F(k))$ to $y = (t + \phi^{-1})^{-1}x$ in the square lattice.

1.28 In the regular lattice of Figure 1.9, show that the area of the parallelogram determined by 0, $F(k + 1)$, $F(k)$, and $F(k - 1)$ is the same for every k, and that the straight line from 0 to $F(k)$ meets the base at the distance $F(k - 2)/F(k)$ from the left point 1.

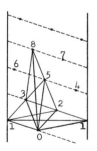

Figure 1.9 Cylindrical lattice built with $d = \phi^{-2}$ (the horizontal distance between point 0, and point 1 on the left), whose base of unit length is determined by two representations of point 1.

2

THE CYLINDRICAL
LATTICE IN BOTANY

Phyllotaxis is a test case in morphogenesis; its empirical and theoretical study deserves to be pursued.

2.1 PRESENTATION

We have already presented a few elementary properties of the cylindrical lattice for a study of the part of botany known as phyllotaxis, which is concerned with the patterns displayed by the similar parts of growing plants. The geometric analysis of this system of references must now be deepened, by means of an extensive use of continued fractions, to deal adequately with the models formulated in it. We now refine the vocabulary, give the complete solution of the problem of the relation between the divergence angles and the visible opposed parastichy pairs (Section 2.3.1), show how the phyllotaxis of a system is related to the continued fraction of its divergence (Sections 2.3.2 and 2.4.1), and give a preview of the modeling work performed in the cylindrical lattice, to explain the origin of phyllotaxis. Section 2.5, entitled Research Activities, presents suggestions of mini-reports or lectures as an incentive to read some of the material regarding phyllotaxis in its original form.

The French botanists Bravais and Bravais originated the cylindrical representation of phyllotaxis. They professed that "In order to deal rigorously with the question of the symmetry of multiple spirals, one must make the following three assertions: (1) the leaves are inserted on a cylinder; (2) the parastichies are geometric helices; and (3) these helices are all parallel and equidistant. Imagine that the cylinder is unfolded in the plane..." (1837, p. 46). Bravais and Bravais suggested (1837, p. 73), long before Turing (1952), that the variation of r can allow us to understand the discontinuous phenomenon of rising phyllotaxis depicted in Section 2.2.2. They have shown that continued fractions have a role to play in such a description (1837, pp. 53, 70).

Most of the ideas contained in Chapter 2 can be found in specialized periodicals. We have organized, polished, and related them so that they can be taught and applied. To their creators we owe a debt of gratitude, particularly to Dr. Irving Adler whose important breakthrough in phyllotaxis pervades this chapter.

2.2 FUNDAMENTAL CONCEPTS

2.2.1 Visible Opposed Parastichy Pair

For most plants the internode distance is not constant; it decreases nearly exponentially as the tip of the apex is reached. But by means of a logarithmic transformation, for example, it is possible to make this distance, that is, the *RISE r*, constant. The *DIVERGENCE d* is constant also; that is, the lattice is *REGULAR*. It is already known that when a leaf is joined to any other one in the *NORMALIZED CYLINDRICAL LATTICE*, a *PARASTICHY* is determined, belonging to a family of parallel, equidistant parastichies, partitioning the set of leaves in pairwise disjoint subsets. Of course, a *LEAF*, or node, is the point or the circular region on the cylinder or on the cylindrical lattice where the (real) leaf, the floret, the scale, or the primordium is inserted. The number n of parastichies in a family is called the *SECONDARY NUMBER* of that family, and a member in the family is an *n-PARASTICHY* (after Bravais and Bravais, 1837). The regular distance between two consecutive leaves on any parastichy in a given family is a *STEP*. If m is a leaf at one step from leaf 0 on an n-parastichy, this parastichy also contains the leaves $2m, 3m, 4m, \ldots$, and it is parallel to the parastichy containing the leaves $1, m + 1, 2m + 1, 3m + 1, \ldots$, and to the parastichy containing the leaves $2, m + 2, 2m + 2, 3m + 2, \ldots$, and so on, up to $m - 1$. Thus, to cover the whose lattice, m parastichies are needed, so that $m = n$. This is *BRAVAIS' THEOREM* used in Section 1.3.1.

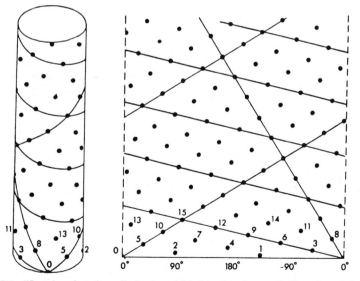

Figure 2.1 "Imagine that the cylinder is unfolded in the plane..." (Bravais and Bravais, 1837, p. 46). Illustration (from Erickson, 1973) of basic concepts presented in the text. (Copyright 1973 by the American Association for the Advancement of Science.)

Two parastichies are said to be *OPPOSED* when they twist around the cylinder in opposite directions. A family of *m*-parastichies winding from the right up to the left together with a family of *n*-parastichies from the left up to the right is an *OPPOSED PARASTICHY PAIR*. It is denoted by (m, n). A *VISIBLE OPPOSED PARASTICHY PAIR* (Adler, 1974) is an opposed pair such that a leaf is at the intersection of any two parastichies. Unless otherwise stated it is supposed that *m* and *n* are relatively prime numbers (see Research Activity 5b in Section 2.5).

Figure 2.1 shows a regular lattice of leaves on a cylinder and the unfolding of its surface in the plane. The straight lines are parastichies, representing the spiral lines of the cylinder. The divergence angle, usually noted as a number smaller than $\frac{1}{2}$, is expressed in degrees. The *GENETIC SPIRAL*, that is, the 1-parastichy passing by the leaves $1, 2, 3, 4, 5, \ldots$, goes from the right up to the left. The two parallel lines going up from left to right are in fact the same 5-parastichy. There are five such parastichies. The opposed pairs $(8, 13)$, $(8, 5)$, and $(3, 5)$ are visible. The pairs $(11, 13)$, $(3, 13)$, and $(11, 5)$ are opposed but not visible. The leaves 3, 5, and 8 are neighbors of the origin.

2.2.2   Conspicuous Parastichy Pair, Point of Close Return

Recall that leaf *n* is a *NEIGHBOR OF THE ORIGIN* at time *T*, if for any other leaf $p \neq 0$, $d(0, p) \geq d(0, n)$, where $d(0, x)$ is the Euclidean distance from leaf 0 to leaf *x*. A *CONTACT* or *CONSPICUOUS PARASTICHY PAIR* at time *T* (Adler, 1977a) is the parastichy pair determined by the first two neighbors *m* and *n* of the origin at that time: $d(0, p) \geq d(0, m) \geq d(0, n)$ for every $p \neq 0$, $p \neq m$, $p \neq n$, and if $d(0, p) = d(0, m)$, $p > m$ and $p > n$. This pair of parastichies determines the *PHYLLOTAXIS*, denoted m/n, $m > n$, of the system. We show in Section 2.4 that a conspicuous pair is necessarily an opposed pair. Figure 2.1 shows 8/5 phyllotaxis.

The phyllotaxis of a system is said to be *NORMAL* if the secondary numbers are consecutive terms in the sequence

$$\langle t, t + 1, 2t + 1, 3t + 2, 5t + 3, \ldots \rangle, \qquad t \geq 2, \qquad (2.1)$$

whose general term is $S_{t, k} = F(k)t + F(k - 1)$, $k = 1, 2, 3, \ldots$, where $F(k)$ is the *k*th term of the main series $1, 1, 2, 3, 5, 8, \ldots$. The phyllotaxis of a system is *ANOMALOUS* for sequences defined by the same recurrence relation as in Expression 2.1, such as the sequence

$$\langle 2, 2t + 1, 2t + 3, 4t + 4, 6t + 7, \ldots \rangle, \qquad t \geq 2, \qquad (2.2)$$

whose general term is $2F(k - 1)t + F(k) + F(k - 2)$, $k \geq 1$, $F(0) = 0$, $F(-1) = 1$, and the sequence

$$\langle t, 2t + 1, 3t + 1, 5t + 2, 8t + 3, \ldots \rangle, \qquad t \geq 2,$$

whose general term is $F(k + 1)t + F(k - 1)$, $k \geq 1$.

RISING PHYLLOTAXIS—a term introduced by Church (1920)—is the passage from one phyllotaxis to another with higher secondary numbers in a growing system. More precisely, it is the passage from m/n to $(m + n)/m$ phyllotaxis. This corresponds to a phenomenon observed in plants. Figure 2.2 represents a sunflower showing two changes in the spiral lines: the

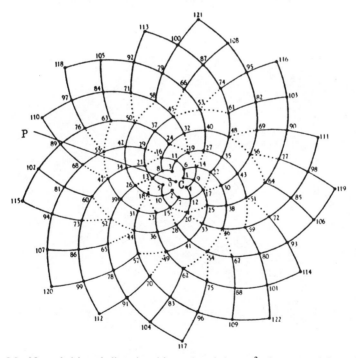

Figure 2.2 Normal rising phyllotaxis, with $t = 2$ and $d = \phi^{-2}$; the terms of the main series $\langle F(k) \rangle$ get closer to $\overline{PC}$ as k increases; the opposed parastichies intersect at 90° (from Richards, 1948).

central zone has 8/5 phyllotaxis, the middle zone has 13/8 phyllotaxis, and the outer zone 21/13 phyllotaxis. Figure 2.3 shows the phenomenon as observed on the capitulum of a sunflower. Williams (1975, pp. 56–81) reconstructed and abundantly illustrated the rising phyllotaxis of *Linum usitatissimum*, going from 1/1 to 2/1, 3/2, 5/3, and 8/5.

The coordinates of leaf n are denoted by

$$(x_n, nr) = (nd - q, nr) \tag{2.3}$$

where q is the integer nearest to nd; x_n is called the *SECONDARY DIVERGENCE* of leaf n (Bravais and Bravais, 1837). We put $D_n = |x_n|$. The *POINTS OF CLOSE RETURN* (Adler, 1974) are the leaves n_i, $i = 1, 2, 3, \ldots$, defined inductively as follows: $n_1 = 1$, and n_{i+1} is the first leaf after n_i such that $D_{n_{i+1}} < D_{n_i}$.

If d is known to be in a given interval, it is possible, as shown in the following simple and direct algorithm, to determine some points of close return.

Figure 2.3   Contact parastichies and rising phyllotaxis of the sunflower; middle zone 21/13, outer zone 34/21.

Algorithm to Determine Points of Close Return. If $1/3 < d < 1/2$, then $2/3 < 2d < 1$, $1 < 3d < 3/2$, $D_2 = 1 - 2d < 1/3 < D_1$, and $n_2 = 2$; we have $n_3 = 3$ only if $D_3 = 3d - 1 < 1 - 2d$, that is, if $d < 2/5$ (leaf 3 is said to be the next close return candidate after $n_2 = 2$). If $1/3 < d < 2/5$, then $n_1 = 1$, $n_2 = 2$, $n_3 = 3$, $2/3 < 2d < 4/5$, $1 < 3d < 6/5$, $4/3 < 4d < 8/5$, $5/3 < 5d < 2$; it follows that $n_4 \neq 4$ (leaf 4 cannot be a point of close return since the possible values for D_4, that is, $4d - 1$ and $2 - 4d$, are greater than and equal to D_2, respectively, and greater than D_3), and we have $n_4 = 5$ only if $2 - 5d < 3d - 1$, that is, if $d > 3/8$ (leaf 5 is the next close return candidate after $n_3 = 3$; see the following definition). The smaller the interval for d is, the greater number of points of close return it is possible to determine.

If n_i is a point of close return for an interval I of values of d, and $n > n_i$ has the property that (1) $n = n_{i+1}$ for a subinterval of I, and (2) for $n_i < b < n$, $D_b \geq D_{n_i}$ in I, then n is called the *NEXT CLOSE RETURN CANDIDATE* after n_i (Adler, 1974). It is denoted by n'_{i+1}.

The rise r is considered as a monotonic decreasing function of the time T, tending toward zero as T tends toward infinity. The leaves present at time T are $0, 1, 2, \ldots, [T]$, since at each unit of time, called a *PLASTOCHRONE*, an actual leaf is born.

2.2.3 Adler's Mediant Nests of Intervals

The notion of mediant between two fractions allows us to understand better the distribution and the law of formation of the intermediate convergents. Those convergents, as we know from Chapter 1, play a significant role in phyllotaxis. The *MEDIANT* between the positive fractions a/b and c/d is defined as the fraction $(a + c)/(b + d)$, a number between a/b and c/d. Furthermore, it is assumed that a and b, like c and d, are relatively prime numbers. For convenience we consider that $0 = 0/1$ and $\infty = 1/0$, so that the mediant between $n = n/1$ and ∞ is $n + 1$, and the mediant between n and $n + 1/m$ is $n + 1/(m + 1)$, where n is a nonnegative integer.

A *MEDIANT NEST OF INTERVALS* $\langle I_k \rangle$, $k = 0, 1, 2, \ldots$, is a sequence of closed intervals defined inductively as follows: $I_0 = [0/1, 1/0]$, and if $I_n = [r, s]$, $n > 0$, I_{n+1} is either $[r, m]$ or $[m, s]$ where m is the mediant between r and s. A mediant nest can be represented by an infinite sequence of ones and zeroes $b_1 b_2 b_3 \cdots$, where $b_{n+1} = 0$ if $I_{n+1} = [r, m]$, and $b_{n+1} = 1$

if $I_{n+1} = [m, s]$, $n > 0$. It can further be represented by $(a_0, a_1, a_2, \ldots)$, since $b_1 b_2 b_3 \cdots$ is made of a_0 consecutive ones, a_1 consecutive zeroes, followed by a_2 consecutive ones, ...

Examples. $1/2 = 001111 \ldots = (0, 2, \infty) = (0, 1, 1, \infty)$; $0 = (0, \infty)$; $\infty = (\infty)$; $6 = (6, \infty) = (5, 1, \infty)$; $\phi = (\sqrt{5} + 1)/2 = (1, 1, 1, 1, \ldots)$. By the well-known principle of closed nested intervals whose lengths tend toward zero, each mediant nest represents a single real number.

Theorem (Adler, 1977a). $\omega = (a_0, a_1, a_2, \ldots)$ if and only if $\omega = [a_0; a_1, a_2, \ldots]$; $\omega = (a_0, a_1, \ldots, a_n, \infty)$ if and only if $\omega = [a_0; a_1, a_2, \ldots, a_n]$.

Proof. The first few intervals of $\omega = (a_0, a_1, a_2, \ldots)$ are $[0/1, 1/0]$, $[1/1, 1/0]$, $[2/1, 1/0], \ldots, [a_0/1, 1/0], [a_0/1, (a_0 + 1)/1], [a_0/1, (2a_0 + 1)/2], \ldots, [a_0/1, (a_1 a_0 + 1)/a_1], \ldots$. Thus after a_0 steps from $[0, \infty]$ we get $[p_0/q_0, p_{-1}/q_{-1}]$, where $p_{-1} = 1$, $q_{-1} = 0$, $p_0 = a_0$ and $q_0 = 1$. After $a_0 + a_1$ steps we obtain the interval $[p_0/q_0, p_1/q_1]$, whose end points are principal convergents of $[a_0; a_1, a_2, \ldots]$. It is easily shown, using induction, that the interval determined after $a_0 + a_1 + \cdots + a_n$ steps is, for $n \geq 1$ and odd, $[p_{n-1}/q_{n-1}, p_n/q_n]$, and for $n \geq 0$ and even, $[p_n/q_n, p_{n-1}/q_{n-1}]$. From the theory of continued fractions (see Research Activity 1f in Section 2.5), this mediant nest of intervals represents the point of convergence of the sequence of principal convergents of $[a_0; a_1, a_2, a_3, \ldots]$.

Now if $\omega = (a_0, a_1, a_2, \ldots, a_n, \infty)$, the interval obtained from $[0, \infty]$ after $a_0 + a_1 + a_2 + \cdots + a_n + m$ steps is $[(mp_n + p_{n-1})/(mq_n + q_{n-1}), p_n/q_n]$ or $[p_n/q_n, (mp_n + p_{n-1})/(mq_n + q_{n-1})]$. Since $\lim_{m \to \infty}(mp_n + p_{n-1})/(mq_n + q_{n-1}) = p_n/q_n$, it follows that $\omega = p_n/q_n = [a_0; a_1, a_2, \ldots, a_n]$.

The converses are true because of the bijection between the set of mediant nests and the set of continued fractions. ∎

Algorithm for Calculating the Convergents of ω from the Mediant Nest of ω. The proof of the theorem yields a very simple algorithm for calculating the principal and intermediate convergents of ω: they are the end points of the intervals of the mediant nest of ω. For example, if $\omega = (0, 1, 2, 3, \infty)$, the consecutive intervals are $[0/1, 1/0]$; $[0/1, 1/1]$; $[1/2, 1/1]$, $[2/3, 1/1]$; $[2/3, 3/4]$, $[2/3, 5/7]$, $[2/3, 7/10]$; etc. The number represented is $7/10$ and

$p_0/q_0 = 0/1, p_1/q_1 = 1/1, p_{2,1}/q_{2,1} = 1/2, p_2/q_2 = 2/3, p_{3,1}/q_{3,1} = 3/4,$ $p_{3,2}/q_{3,2} = 5/7$, $p_3/q_3 = 7/10$, and so on. In the sequence of nested intervals, the semicolon is used to indicate that a principal convergent has appeared as an end point of an interval.

Exercises

2.1 Suppose that leaf n is a neighbor of the origin. Show that a leaf $m > n$ cannot become a neighbor of the origin as r decreases, unless $D_m < D_n$.

2.2 Given $2/5 < d < 1/2$, show that 5 is the next close return candidate after 2 only if $d < 3/7$.

2.3 Show that in the interval $3/8 < d < 5/13$, normal phylotaxis with $t = 2$ is possible.

2.4 Show that in the interval $5/12 < d < 8/19$, anomalous rising phyllotaxis with $t = 2$ is possible.

2.5 The symbolism used in this exercise on continued fractions is fully explained in Chapter 1. Let $\omega = [a_0; a_1, a_2, a_3, \ldots]$, where $a_n > 1$ for $n \geq 2$. Show that the sequence

$$\frac{p_{n-2}}{q_{n-2}}, \frac{p_{n,1}}{q_{n,1}}, \frac{p_{n,2}}{q_{n,2}}, \ldots, \frac{p_{n,c}}{q_{n,c}}, \ldots, \frac{p_n}{q_n}$$

decreases when n is odd and increases when n is even. Show that in the first case, each term is greater than p_{n-1}/q_{n-1}, and that in the other case each term is smaller than p_{n-1}/q_{n-1} (compare with Exercises 1.12 and 1.23). Devise an algorithm to determine p_n/q_n and a_n from the values of ω, p_{n-2}/q_{n-2}, and p_{n-1}/q_{n-1}.

2.6 Determine all the convergents, principal and intermediate, of $\omega = [1; 2, 1, 2, \ldots]$ from the mediant nest $(1, 2, 1, 2, \ldots)$.

2.7 Determine the value of $\omega = [1; 2, 3]$ from $(1, 2, 3, \infty)$.

2.8 Show that if $\omega = (a_0, a_1, a_2, \ldots)$, $a_0 \leq \omega \leq a_0 + 1$ and the sequence of a_i's terminates if and only if ω is rational or ∞. Letting $a_n > 1$, show that $\omega = (a_0, a_1, a_2, \ldots, a_{n-1}, a_n, \infty)$ if and only if $\omega = (a_0, a_1, a_2, \ldots, a_{n-1}, a_n - 1, 1, \infty)$; every positive rational number possesses exactly two mediant nests.

2.3 DIVERGENCE ANGLES VERSUS VISIBLE OPPOSED PARASTICHY PAIRS

2.3.1 Contractions and Extensions of Parastichy Triangles

Let (m, n) be a visible opposed parastichy pair and a and A two representations of leaf 0 at $(0, 0)$ and $(1, 0)$, respectively, in the normalized cylindrical lattice. Let S be the leaf mn. Triangle aSA is by definition the *VISIBLE OPPOSED PARASTICHY TRIANGLE* belonging to the pair (m, n) (Adler, 1974). It is denoted by $\Delta(m, n)$. Figure 2.1 shows $\Delta(3, 5)$ and $\Delta(8, 5)$; $\Delta(3, 2)$ and $\Delta(2, 1)$ can easily be drawn. The first leaf on the left side of $\Delta(m, n)$ is leaf n; leaf mn is at m steps from a, and at n steps from A. If $m > n$, let V be the leaf on $\overline{aS}$ at n steps from S. Triangle aVA, illustrated in Figure 2.4, is by definition the *CONTRACTION OF* $\Delta(m, n)$, and $(m - n, n)$ is the *CONTRACTION OF* (m, n). In Figure 2.1, $\Delta(3, 5)$ is the contraction of $\Delta(8, 5)$. If $m < n$, point V is at m steps from S on $\overline{AS}$. Obviously the contraction of (m, n) is a visible pair.

Now let V' be the leaf on the extension of $\overline{AS}$ at m steps from S and V'' be the leaf on the extension of $\overline{aS}$ at n steps from S, as in Figure 2.4. The

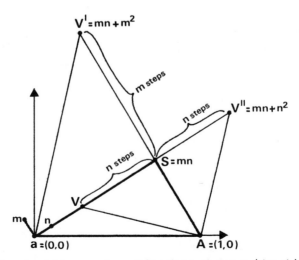

Figure 2.4 Given the visible opposed parastichy pair (m, n), $m > n$, $\Delta(m, n)$ is the triangle aSA, aVA is its contraction, and $aV'A$ and $aV''A$ are its right and left extensions, respectively.

triangles $aV'A$ and $aV''A$ are, respectively, the *RIGHT* and the *LEFT EXTENSIONS OF* $\Delta(m, n)$; the right and the left sides of $\Delta(m, n)$, respectively, have been extended. The pair determined by the $(m + n)$-parastichies and the n-, or the m-, parastichies is not necessarily opposed, but when it is opposed, it is visible. The right extension corresponds to the pair $(m, m + n)$, and the left extension to the pair $(m + n, n)$. If a visible opposed parastichy pair (m, n) is the result of $k = 0, 1, 2, 3, \ldots$ consecutive extensions of the visible pair (p, q), then, by definition, (m, n) is an *EXTENSION OF ORDER k* of $(p, q), (p, q)$ being the extension of order 0. If the sequence of extensions of (p, q) is made, alternately, of right and left extensions, (m, n) is called an *ALTERNATE EXTENSION* of (p, q).

Fundamental Theorem on the Visibility of Extensions (Adler, 1974).   If the closed interval $[x/y, z/w]$, where $0 < x/y < z/w \leq \frac{1}{2}$, and the opposed parastichy triangle aSA, illustrated in Figure 2.5, have the properties that (1) the left side of the triangle has w steps, each step having a projection on $\overline{aA}$ of length equal to $-x + yd$, and the right side of the triangle has y steps, each one having a projection on $\overline{aA}$ of length $z - wd$, and (2) aSA is a visible opposed parastichy triangle if and only if $x/y \leq d \leq z/w$, then the

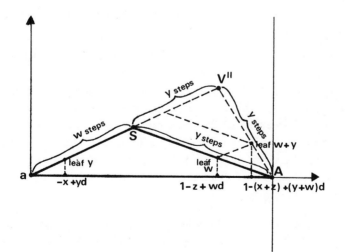

Figure 2.5   Illustration of the fundamental theorem on the visibility of extensions, for the left extension $aV''A$ of aSA.

interval $[x/y, (x + z)/(y + w)]$ and the left extension of aSA have the same properties, and the interval $[(x + z)/(y + w), z/w]$ and the right extension of aSA have the same properties.

Proof. It must be proved for the left extension $aV''A$ (Figure 2.5) that (1) the left side $\overline{aV''}$ has $y + w$ steps, (2) each of these steps has a projection of length $-x + yd$, (3) the right side has y steps, (4) each of these has a projection of length $(x + z) - (y + w)d$, and finally (5) $aV''A$ is a visible opposed parastichy triangle if and only if $x/y \le d \le (x + z)/(y + w)$. Consider part 4 of the problem. We are looking for the abscissa of leaf $w + y$, that is, for the value of $(w + y)d - [(w + y)d]$. By hypothesis, $wd - [wd] = 1 - z + wd$, and $yd - [yd] = -x + yd$, where $[\cdot]$ means the integral part. We then have $[wd] = z - 1$, $[yd] = x$, $x \le yd < x + 1$, $z - 1 \le wd < z$, $x + z - 1 \le yd + wd < x + z + 1$, $[(y + w)d] = x + z - 1$ or $x + z$, and finally the abscissa we are looking for has the value $(w + y)d - (x + y) + 1$ or $(w + y)d - (x + z)$. This last value is smaller than $-x + yd$, meaning that the $(w + y)$-parastichies have the same direction as the y-parastichies, and that the triangle $aV''A$ is not an opposed parastichy triangle. The other value means that extension $aV''A$ is an opposed parastichy triangle, and that $d \le (x + z)/(y + w)$ (part 5). In both cases the length of the projection is $(x + z) - (y + w)d$. When the parastichies are not opposed, the abscissa of $y + w$ is $(w + y)d - (x + z) > 0$, in which case $d > (x + z)/(y + w)$. Part 5 is thus proved. For the right extension the proof is similar. ∎

Notice that $(t, t + 1)$ is a visible opposed parastichy pair if and only if $1/(t + 1) \le d \le 1/t$ (see Exercises 2.9–2.11).

Here is an important result in botanometry. It was partly produced in Section 1.3.2, for neighbors of the origin, that is, for conspicuous parastichy pairs, in the framework of Bravais's approximation formula. The result is a straightforward corollary of Exercises 2.9 and 2.12 and of the fundamental theorem on the visibility of extensions, but it is most easily obtained from Exercises 1.16 and 1.17. The case $t = 2$ gives the main series and nested intervals containing ϕ^{-2}.

Adler's Theorem (1974).   Let (m, n) be a visible opposed parastichy pair. Then n and m are the terms $S_{t, k}$ and $S_{t, k+1}$ of the normal phyllotaxis sequence, if and only if d is equal to $F(k)/S_{t, k}$ or $F(k + 1)/S_{t, k+1}$ or is between these two values; at the limit $d = (t + \phi^{-1})^{-1}$.

We show, in Section 2.4 and in Exercises 2.20, 2.27, and 2.33, that if m and n are points of close return and a fortiori, if m/n is the phyllotaxis of the system, the interval of values for d is shorter and included in the interval given in the preceding theorem.

2.3.2 Visible Pairs versus Continued Fraction of the Divergence

The next two algorithms, useful by themselves, will help us to understand the facts stated in the proposition of this section.

Determining the Visible Pairs from the Value of d. Suppose that $d = 3/16$. The continued fraction of d is $[0; 5, 3]$ and d has two mediant nests, that is, $(0, 5, 3, \infty)$ and $(0, 5, 2, 1, \infty)$. The first few intervals in the first nest are $[0/1, 1/0]$; $[0/1, 1/1]$, $[0/1, 1/2]$, $[0/1, 1/3]$, $[0/1, 1/4]$, $[0/1, 1/5]$; $[1/6, 1/5]$, $[2/11, 1/5]$, $[3/16, 1/5]$; ...; in the second nest we have the same intervals up to $[2/11, 1/5]$, followed by $[2/11, 3/16]$; .... In both cases this means that $(5, 6)$ is a visible pair since $1/6 \le d \le 1/5$. Thus the contractions of $(5, 6)$, that is, $(5, 1)$, $(4, 1)$, $(3, 1)$, $(2, 1)$, and $(1, 1)$, are visible. The visible extensions corresponding to the first mediant nest are $(5, 11)$, $(5, 16)$; $(21, 16)$, $(37, 16)$, and all other consecutive left extensions. The visible extensions corresponding to the other nest are $(5, 11)$; $(16, 11)$; $(16, 27)$, $(16, 43)$, ... followed by an infinity of right extensions. In the sequence of visible opposed parastichy pairs, the semicolon is used to indicate a change in the direction of the extensions.

Determining an Interval for d from a Visible Pair. Suppose that the pair $(21, 58)$ is visible. Let us determine the consecutive contractions of the pair, up to $(0, 1)$, in order to obtain a sequence of visible extensions of $(0, 1)$ up to $(21, 58)$, that is, $(0, 1)$; $(1, 1)$, $(2, 1)$, $(3, 1)$, $(4, 1)$, $(5, 1)$; $(5, 6)$, $(5, 11)$, $(5, 16)$; $(21, 16)$; $(21, 37)$, $(21, 58)$. Again, the semicolon indicates a change in the direction of the extensions. This means that $d = (0, 5, 3, 1, 2 + x, ...)$ (that is, $a_0 = 0$, $a_1 = 5$, $a_2 = 3$, $a_3 = 1$, $a_4 \ge 2$, where the a_i are consecutive entries in the continued fraction of d) since the intervals in the mediant nest are $[0/1, 1/0]$; $[0/1, 1/1]$, $[0/1, 1/2]$, $[0/1, 1/3]$, $[0/1, 1/4]$, $[0/1, 1/5]$; $[1/6, 1/5]$, $[2/11, 1/5]$, $[3/16, 1/5]$; $[3/16, 4/21]$; $[7/37, 4/21]$, $[11/58, 4/21]$, ... As seen in Section 2.2.2, the end points of these intervals are convergents of d. More precisely the principal convergents are $p_0/q_0 = 0/1$, $p_1/q_1 = 1/5$, $p_2/q_2 = 3/16$, $p_3/q_3 = 4/21$. We will have $p_4/q_4 =$

$11/58$ if $a_4 = 2$, and $p_4/q_4 = 15/79$ if $a_4 = 3$, and so forth. In all cases we have $11/58 \le d \le 4/21$.

Proposition

Let (m, n) be a visible opposed parastichy pair in a distribution of leaves with divergence d. Then the smallest secondary number, m or n, is the denominator of a principal convergent of d, that is, a neighbor of the vertical axis of the cylindrical lattice. The largest secondary number is the denominator of a principal convergent of d if and only if in the sequence of visible extensions of $(0, 1)$ leading to (m, n), (m, n) immediately precedes a change in the direction of the extensions.

Proof. The mediant nest starting with $[0/1, 1/0]$ and the visible pairs starting with the pseudopair $(0, 1)$, can be put into a one-to-one correspondence. The convergents and the visible pairs can also be put into a one-to-one correspondence. The visible pair that precedes a change in the direction of the extensions corresponds to the appearance of a principal convergent as an end point of an interval of the nest. The smaller of the two numbers, say, n, is necessarily the denominator of a principal convergent. On the other hand, m is the denominator of an intermediate convergent unless there is a change of direction after the appearance of m as a denominator of an end point of an interval of the nest. ■

The following theorem (Adler, 1977a) is a corollary of the preceding proposition and of Exercise 2.19.

Theorem. Let (m, n) be a visible opposed parastichy pair of a distribution of leaves with divergence d, where m and n are denominators of principal convergents of d. Then m and n are denominators of consecutive principal convergents.

Exercises

2.9   If (m, n), m and $n > 1$, is a visible opposed parastichy pair, show that it is an extension of a certain order of a unique visible opposed pair $(t, t + 1)$ or $(t + 1, t)$, $t > 1$. Conclude that $1/(t + 1) \le d \le 1/t$. Finally, show that $(t, t + 1)$ [or $(t + 1, t)$] is a visible opposed parastichy pair if and only if $1/(t + 1) \le d \le 1/t$.

2.10 Show that the left extension $(2t + 1, t + 1)$ of $(t, t + 1)$ is visible if and only if $1/(t + 1) \le d \le 2/(2t + 1)$. Show that the right extension $(t, 2t + 1)$ of $(t, t + 1)$ is visible if and only if $2/(2t + 1) \le d \le 1/t$. Repeat the process.

2.11 Show that $(13, 8)$ is visible if and only if $3/8 \le d \le 5/13$. Show that $(19, 12)$ is visible if and only if $5/12 \le d \le 8/19$. Compare with Exercises 2.3 and 2.4.

2.12 Let (m, n) be an alternate visible extension of order k of $(t, t + 1)$, starting with a left extension. Then show:
(a) If k is even, $m = S_{t, k+1}$, $n = S_{t, k+2}$, and $F(k + 2)/S_{t, k+2} \le d \le F(k + 1)/S_{t, k+1}$.
(b) If k is odd, $m = S_{t, k+2}$, $n = S_{t, k+1}$ and $F(k + 1)/S_{t, k+1} \le d \le F(k + 2)/S_{t, k+2}$.
[When the genetic spiral is from the right up to the left, start with a right extension of $(t + 1, t)$.]

2.13 Show that there exists an infinite sequence of alternate visible extensions of $(t, t + 1)$, starting with a left extension, if and only if $d = (t + \phi^{-1})^{-1}$.

2.14 State and prove a result similar to Adler's theorem of Section 2.3.1, for the sequence given in Expression 2.2, and for the sequence $\langle F(k + 1)t + F(k - 1)\rangle$, $k = 1, 2, 3, \ldots, t \ge 2$.

2.15 If $d = 5/12$ determine two infinite sequences of visible opposed parastichy pairs.

2.16 If $d = \sqrt{3}/6$, show that the sequence of visible opposed parastichy pairs is $(1, 1)$, $(2, 1)$, $(3, 1)$, $(3, 4)$, $(3, 7)$, $(10, 7)$, $(17, 7)$, $(24, 7)$, $(31, 7)$, $(38, 7)$, $(45, 7), \ldots$.

2.17 If the pair $(17, 39)$ in a distribution of leaves is visible, determine the possible values for d. Determine the first few principal convergents of d.

2.18 If $d = [0; 4, 2, 1, 2, 1, 2, 1, \ldots]$ calculate the first few convergents and determine the intervals in the mediant nest and the first few visible pairs.

2.19 Let (m, n) be a visible opposed parastichy pair, in a regular distribution of leaves, which immediately precedes a change in the direction of the visible extensions of $(0, 1)$ [leading to (m, n)]. Then show that m and n are denominators of consecutive principal convergents of d.

2.4 PHYLLOTAXIS OF A SYSTEM

2.4.1 Adler's Theorem on Points of Close Return

Let Z be the set of neighbors of the origin obtained by the continuous growth of time T, R the set of points of close return, and V the set of neighbors of the vertical axis of the lattice. It is obvious that $Z \subseteq R$ (see Exercise 2.1), and that $R \subseteq V$. But the consecutive points of close return alternate on either side of the vertical axis, and R coincides with the set of principal neighbors of this axis, that is, with the denominators of the principal convergents of the continued fraction of d (See Section 2.5, Research Activity 1i). We have $n_i = q_{i-1}$ and $D_{n_i} = |q_{i-1}d - p_{i-1}|$, $n_{i+1} = q_i$ and $D_{n_{i+1}} = |q_i d - p_i|$, $i = 1, 2, 3, \ldots$. The phyllotaxis of a system is thus determined by denominators m and n of consecutive principal convergents, and the parastichies of the pair (m, n) are opposed. Notice that d determines the set R, the rise r determines which n_i's are neighbors of the origin, and T determines which n_i's exist. The following proposition states that $Z = R$.

Proposition 1

If r is such that $d(0, n_{i+1}) > d(0, n_i)$, for $i = 1, 2, 3, \ldots$, then, as r tends toward 0, the n_i's, in their natural order, become consecutive neighbors of the origin (Adler, 1975).

Proof. Initially $n_1 = 1$ is a neighbor of the origin, and $d(0, n_1) < d(0, n_2)$. Let us suppose that n_i is a neighbor of the origin. It must be shown that n_{i+1} will be a neighbor before n_{i+k}, $k > 1$. Points n_i and n_{i+1} are on different sides of the vertical axis of the lattice. Let us suppose that n_i is in the first quadrant. Point n_{i+k} can be in the first or second quadrant. Let α and β be the angles determined with the X-axis by the perpendicular bisectors of the segments $\overline{n_i n_{i+1}}$ and $\overline{n_i n_{i+k}}$, respectively, meeting the axis at points P and Q, respectively. It is obvious that $90° > \alpha > \beta$ since $n_{i+k} > n_{i+1} > n_i$. As r decreases, point P reaches point 0, the origin, before point Q does so, since Q is on the left of P. This means that $d(0, n_{i+1}) = d(0, n_i)$ before $d(0, n_{i+k}) = d(0, n_i)$. ■

Proposition 2

Let m and n, $m > n$, relatively prime, be two consecutive points of close return, with respective coordinates $(x_m, mr) = (md - p, mr)$ and (x_n, nr)

$= (nd - q, nr)$ (Expression 2.3), of a distribution of leaves with divergence d in the normalized cylindrical lattice. Then we have (compare with Exercise 1.16)

$$np - mq = \pm 1, \qquad (2.4)$$

an equation which has at most two solutions,

$$mD_n + nD_m = 1, \qquad mnr^2 - r \cot \beta + x_n x_m = 0, \qquad (2.5)$$

where β is the angle of intersection of the opposed parastichies, and for $\beta = \pi/2$, $r = (-x_n x_m/mn)^{1/2}$, $r^2 = -d^2 + (mq + np)d/mn - pq/mn$.

Proof. Suppose for definiteness that $m = q_i$ and $n = q_{i-1}$; then $p = p_i$ and $q = p_{i-1}$. The first relation is one that exists between consecutive principal convergents (Exercise 1.7). From indeterminate analysis we know that this relation has integral solutions, but on the condition that m and n are relatively prime. This relation has either no solution or one solution for $+1$, as i is odd or even, respectively. For -1, the equation has either no solution or one solution, as i is even or odd, respectively. Suppose that n is on the left of the vertical axis. This means that $D_m = q_i d - p_i$, $D_n = p_{i-1} - q_{i-1}d$, and i is even, since the convergents of even order are smaller than d (see Section 1.2.3 or Research Activity 1 in Section 2.5). It follows that $mD_n + nD_m = (-1)^i = 1$. Let us suppose now that n is on the right of the axis: i is odd and $mD_n + nD_m = (-1)^{i+1} = 1$. ∎

The following proposition is an immediate corollary of Proposition 2.

Proposition 3

If n, m, and $m + n$ are three consecutive points of close return,

$$D_n = D_m + D_{m+n}, \qquad (2.6)$$

$$x_{m+n} = x_m + x_n. \qquad (2.7)$$

Proposition 4

If m/n, $m > n$, is the phyllotaxis of a system such that $d(0, m) = d(0, n)$, then the pair (m, n) is visible (Adler, 1977a).

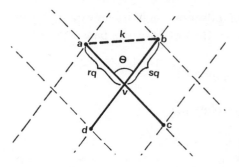

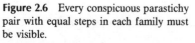

Figure 2.6 Every conspicuous parastichy pair with equal steps in each family must be visible.

Proof. Suppose there is no leaf at the intersection v of an m-parastichy and an n-parastichy. The situation, illustrated in Figure 2.6, is the following: a, b, c, and d are leaves (such as leaves 13, 18, 10, and 5 in the $(3, 13)$ system of Figure 2.1), θ is angle avb, $\overline{ab} = k$, and $\overline{ac} = \overline{bd} = q$. For two of the four leaves, say, a and b, we have $\overline{av} = rq$ and $\overline{bv} = sq$, where r and s are smaller than or equal to $\frac{1}{2}$. By the law of cosines applied to triangle abv we can write

$$k^2 = r^2q^2 + s^2q^2 - 2rsq^2\cos\theta.$$

Since $k \geq q$ (m/n is the phyllotaxis of the system), it follows that

$$1 \leq \frac{k^2}{q^2} = r^2 + s^2 - 2rs\cos\theta < (r + s)^2 \leq 1,$$

an impossible relation. Thus (m, n) is a visible pair. ∎

Under the hypothesis of Proposition 4, given the theorem of Section 2.3.2, if $d = (t + \phi^{-1})^{-1}$ then m and n are denominators of consecutive principal convergents. The following theorem is formulated for intervals of values of d associated with normal phyllotaxis.

Fundamental Theorem on Points of Close Return (Adler, 1974). If $F(2k)/S_{t,2k} < d < F(2k - 1)/S_{t,2k-1}$ for $k \geq 1$, then $n_i = S_{t,i-1}$, $D_{n_i} = (-1)^i(F(i - 1) - dn_i)$ for $i = 1, 2, \ldots, 2k$, and $n'_{2k+1} = S_{t,2k}$, $D_{n'_{2k+1}} = dn'_{2k+1} - F(2k)$.

If $F(2k)/S_{t,2k} < d < F(2k + 1)/S_{t,2k+1}$ for $k \geq 1$, then $n_i = S_{t,i-1}$, $D_{n_i} = (-1)^i(F(i - 1) - dn_i)$ for $i = 1, 2, \ldots, 2k + 1$, and $n'_{2k+2} = S_{t,2k+1}$, $D_{n'_{2k+2}} = F(2k + 2) - n'_{2k+2}d$.

Proof. We concentrate on the first result, leaving the proof of the other as an exercise. For $k = 1$, $1/(t + 1) < d < 1/t$, and, by Exercise 2.9, $(t, t + 1)$ is visible. By taking the consecutive contractions of $(t, t + 1)$ and reversing the process, we can see that the pair $(t, 1)$ immediately precedes a change in the direction of the extensions leading to $(t, t + 1)$. By Exercise 2.19, this means that 1 and t are consecutive points of close return (since the points of close return coincide with the denominators of principal convergents). So n_1 and n_2 are as predicted, that is, equal to $S_{t,0}$ and $S_{t,1}$, respectively. Now $t + 1$ can be a point of close return only if $(t + 1)d - 1 < 1 - td = D_t$, that is, if $d < 2/(2t + 1)$, which is possible. It follows that $n'_3 = t + 1 = S_{t,2}$ and $D_{t+1} = (t + 1)d - 1$. The result is thus true for $k = 1$.

Making an induction, let us suppose that the result is true for a fixed $k > 1$. We must prove that it is true for $k + 1$. If $F(2k + 2)/S_{t,2k+2} < d < F(2k + 1)/S_{t,2k+1}$, then the result is true for $i = 1, 2, \ldots, 2k + 2$ and $n'_{2k+3} = S_{t,2k+2}$, $D_{n'_{2k+3}} = dn'_{2k+3} - F(2k + 2)$. In the given interval, the pair $(S_{t,2k+1}, S_{t,2k+2})$ is visible (see Exercises 2.9 and 2.12a). It follows that n_i, $i = 1, 2, \ldots, 2k + 2$, is a point of close return (see Exercise 2.19). Since $F(2k)/S_{t,2k} < d < F(2k - 1)/S_{t,2k-1}$ (see Exercise 2.20a), the induction hypothesis gives $D_{n_i} = (-1)^i(F(i - 1) - dn_i)$ for $i = 1, 2, \ldots, 2k$. By Proposition 2 applied to the pair n_{2k} and n_{2k+1} and then to the pair n_{2k+1} and n_{2k+2}, using the relation $mD_n + nD_m = 1$ (Expression 2.5) for these pairs of relatively prime numbers, we successively get the values of $D_{n_{2k+1}}$ and $D_{n_{2k+2}}$, as expected.

It thus remains to find n'_{2k+3} and $D_{n'_{2k+3}}$. By multiplying the interval for d by $S_{t,2k+2}$, it follows that $F(2k + 2) < S_{t,2k+2}d < F(2k + 2) + 1/S_{t,2k+1}$ (see Exercise 2.20a), and that $D_{S_{t,2k+2}} = S_{t,2k+2}d - F(2k + 2)$. It is easily verified that this value is smaller than $D_{S_{t,2k+1}}$ in the interval $F(2k + 2)/S_{t,2k+2} < d < F(2k + 3)/S_{t,2k+3}$, where $S_{t,2k+2} = n'_{2k+3}$ is a point of close return. Let b be an integer such that $S_{t,2k+1} < b < S_{t,2k+2}$. It is easily shown that D_b has one of the values $D_b = F(2k + 2) - (S_{t,2k+1} + i)d$, or $D_b = (S_{t,2k+1} + i)d - F(2k + 1)$, where $i = 1, 2, \ldots, S_{t,2k} - 1$, and that $D_b > D_{S_{t,2k+1}}$ in both cases. ∎

As a result of his observations, Wright proposed that $\frac{1}{3} < D_{k+1}/D_{k-1} < \frac{1}{2}$. He wrote indeed (1873, p. 399) that "Each leaf of the cycle is so placed over the space between older leaves nearest in direction to it as always to fall near the middle, and never beyond the middle third of the space, or by more than one sixth of the space from the middle, until the cycle is completed,.... This property depends mathematically on the character of the continued fractions...." This is true, as one can see in Exercise 2.21. This property is in fact characteristic of normal phyllotaxis, expressed in Exercise 2.22. It plays a key role in Richards' explanation (1948, pp. 225–226) of normal phyllotaxis.

Exercise 2.23 presents already known results (Section 1.4.3, Exercises 1.8 and 1.26, and the theorem of Section 1.4.1), dealt with here in a slightly different way, since d is in all the open intervals of the fundamental theorem on points of close return. Exercise 2.24 generalizes Exercise 1.21.

2.4.2   The (d, r) Phase-Space

Since a distribution of leaves is a regular cylindrical lattice, its phyllotaxis, at each time T, is completely determined by the values of d and r, the coordinates of leaf 1. This distribution can then be defined as a point in the two-dimensional phase-space, that is, in the (d, r) plane. The *PHYLLOTACTIC BIOGRAPHY* of a distribution of leaves is represented by a path in the phase-space. This section analyzes the variation of the point (d, r) on arcs of circles determined by neighbors of the origin, equidistant from the origin.

Consider the cylinder of Figure 2.7, with radius equal to 1, on which the centers of leaves 0, 1, and m are inserted; the leaves are spheres of equal

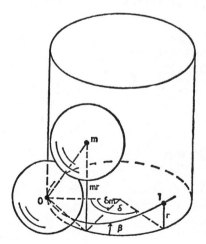

Figure 2.7   Illustration (from Erickson, 1973) of Proposition 1 showing relations between r, d, and the conspicuous pair (m, n). (Copyright 1973 by the American Association for the Advancement of Science.)

radii. Proposition 1 demonstrates the interdependence of the parameters $\delta = 2\pi d$ and r, when m/n is the phyllotaxis of the system and $d(0, m) = d(0, n)$.

Proposition 1 (van Iterson, 1907)

If m/n is the phyllotaxis of a system of circular leaves of equal radii, where the spheres centered at m and n are tangent to the sphere centered at 0 $[d(0, m) = d(0, n)]$ and $\delta = 2\pi d$, then we have

$$r^2(m^2 - n^2) + 4\sin\left(\frac{(m + n)\delta}{2}\right)\sin\left(\frac{(m - n)\delta}{2}\right) = 0. \qquad (2.8)$$

If, moreover, the sphere centered at $m + n$ is tangent to the one centered at 0 $[d(0, m) = d(0, m + n)]$, then

$$\frac{m^2 - n^2}{n(n + 2m)} = \frac{\sin[(m + n)\delta/2]\sin[(m - n)\delta/2]}{\sin[(n + 2m)\delta/2]\sin(n\delta/2)}. \qquad (2.9)$$

Proof. Since the lengths of the chords subtending the secondary divergence angles $\delta_m = m\delta - 2\pi p$ and $\delta_n = n\delta - 2\pi q$ are equal to $2\sin(\delta_m/2)$ and $2\sin(\delta_n/2)$, respectively, where p and q are the integers such that $2\pi p$ and $2\pi q$ are nearest to $m\delta$ and $n\delta$, respectively, we have

$$r^2(m^2 - n^2) + 4\sin^2\left(\frac{m\delta}{2}\right) - 4\sin^2\left(\frac{n\delta}{2}\right) = 0,$$

and the results follow immediately. ∎

In general it is not possible to solve Expression 2.9 directly; numerical analysis is needed. However, when $n = 1$ and $m = 2$ we get $\delta \simeq 2.3005$, that is, $d \simeq 0.366$. The method of arcs (n_{i+1}, n_i), in the normalized cylinder unrolled in the plane, proposed in Propositions 2–6 of this section, rapidly gives $\delta = 5\pi/7$; that is, $d = \frac{5}{14} \simeq 0.357$ [d represents the point of intersection of arc $(2, 1)$ with arc $(3, 2)$]. The difference between the two values of δ comes from the fact that a chord and an arc respectively are considered. In the last case the right side of Expression 2.9 must be replaced by $(\delta_n^2 - \delta_m^2)/(\delta_m^2 - \delta_{m+n}^2)$, and we must use Expressions 2.4 and 2.7 to obtain the value of δ (Expression 2.4 is not modified when the circumference of the base of the cylinder is 2π).

Exercises 2.25 and 2.26 illustrate the observation that in normal rising phyllotaxis with $t = 2$, the rhombus made by leaves n, m and $m + n$ equidistant from leaf 0 at the origin of coordinates is repeated by leaves 0, m, $m + n$, and $2m + n$ and shows an angle of 120°.

When m/n is the phyllotaxis of a system such that $d(0, m) = d(0, n)$, Proposition 2 shows that d is a quadratic function of $d(0, m)$. Figure 2.8 gives an idea of this dependence for a few types of phyllotaxis and of the relation between the rising phyllotaxis of a system and the sequence of nested intervals of values for d (see Exercises 2.27, 2.29, and 2.33).

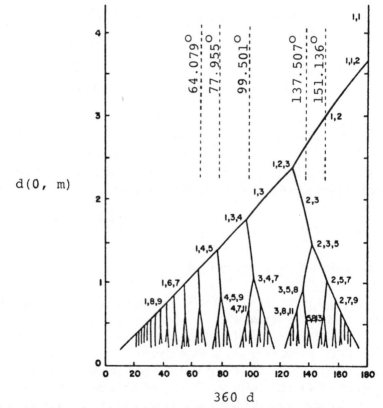

Figure 2.8 Adapted version of Veen and Lindenmayer's design (1977) of van Iterson's graphical work (1907) for the arcs of parabolas given by Expression 2.10, for a few values of m and n.

Proposition 2 (van Iterson, 1907)

Given a system whose phyllotaxis is m/n, and where $d(0, m) = d(0, n)$,

$$d = \epsilon \frac{\left[(m^2 - n^2)d^2(0, m) + 1\right]}{2nm} + \frac{\Delta_n}{n} \tag{2.10}$$

where Δ_n is an integer depending on n, and $\epsilon = \pm 1$.

Proof. $d^2(0, n) = (nd - q)^2 + n^2r^2$ and $d^2(0, m) = (md - p)^2 + m^2r^2$, where p and q are the integers nearest to md and nd, respectively. Extracting the value of r^2 from the second equation yields the result, where $\Delta_n = q$, by Expression 2.4 (q is sometimes called the *ENCLYCLIC NUMBER* of n).

∎

Given a distribution of leaves with divergence angle d, with n_{i+1} and n_i two consecutive points of close return, let us consider the functions of d, $f(d) = d^2(0, n_{i+1})$ and $g(d) = d^2(0, n_i)$. The following three propositions on these functions can be very easily demonstrated.

Proposition 3

Functions f and g are arcs of parabolas whose vertices are respectively given by $(p_i/q_i, q_i^2 r^2)$ and $(p_{i-1}/q_{i-1}, q_{i-1}^2 r^2)$. The values $d = p_i/q_i$ and $d = p_{i-1}/q_{i-1}$ minimize f and g, respectively, and $f(p_{i-1}/q_{i-1}) > g(p_{i-1}/q_{i-1})$. The condition $f(p_i/q_i) < g(p_i/q_i)$ means that the parabolas have a common point ($f = g$), and every value of r such that

$$r < \frac{1}{q_i\left(q_i^2 - q_{i-1}^2\right)^{1/2}} \tag{2.11}$$

realizes the condition.

Proposition 4

The condition $f = g$, when possible, defines an arc of a circle, in the first quadrant of the (d, r) plane, whose center and radius are respectively given by

$$\left(\frac{p_i q_i - p_{i-1} q_{i-1}}{q_i^2 - q_{i-1}^2}, 0 \right), \tag{2.12}$$

$$\frac{1}{q_i^2 - q_{i-1}^2}. \tag{2.13}$$

The arc of a circle defined in Proposition 4 is called $ARC\ (n_{i+1}, n_i)$ [or arc (n_i, n_{i+1}), depending on the direction of the genetic spiral]. The length of this arc is well defined, since n_i and n_{i+1} are consecutive points of close return only in a given interval of values of d (see Exercise 2.29).

Proposition 5

If r, fixed, satisfies the condition given in Proposition 3, and if n_{i+1} is a neighbor of the origin, then the value of d which maximizes $d(0, n_{i+1})$ is on the arc (n_{i+1}, n_i).

Figure 2.9 summarizes the data in Exercises 2.28–2.30. These are very simple exercises regarding the four arcs (n_{i+1}, n_i) (in heavy lines in the

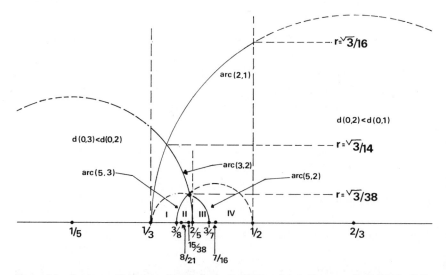

Figure 2.9 Summary of relevant data in the region $1/3 < d < 1/2$ on arcs $(2,1)$, $(3,2)$, $(5,2)$, and $(5,3)$.

figure) which are relevant when $T \leq 5$ and $1/3 < d < 1/2$. A figure such as Figure 2.8 can be drawn with arcs of circles instead of arcs of parabolas. For the needs of Proposition 6, it must be noticed that regions I, II, III, and IV in Figure 2.9 present the following relations:

$$\text{I:} \quad d(0,3) < d(0,5) < d(0,2),$$
$$\text{II:} \quad d(0,5) < d(0,3) < d(0,2),$$
$$\text{III:} \quad d(0,5) < d(0,2) < d(0,3),$$
$$\text{IV:} \quad d(0,2) < d(0,5) < d(0,3).$$

Proposition 6 underlines fine properties of the cylindrical lattice that are crucial in the model of Section 5.2, where Adler shows that a sufficient condition for ϕ^{-2} to appear is that when contact pressure begins $r \geq \sqrt{3}/38$ or $T < 5$. In the proposition, the expression $\min d(0, n)$ represents the minimal distance between the leaves of the lattice, more precisely between leaf 0 and the point of close return n, neighbor of 0. By Proposition 5, this minimal distance is maximized on the arc (n_{i+1}, n_i). Contact pressure corresponds precisely to the mathematical statement that $\min d(0, n)$ is maximized, that is, to a value below which the distance from leaf 0 to its nearest neighbor begins to decrease. Under the conditions stated in the above main result of Adler's model, by Proposition 6 we have 2/1 or 3/2 phyllotaxis, which rises, as can be proved, along the main series $\langle F(k) \rangle$ if contact pressure is maintained (see Exercise 2.34). For $T = 5$ and $r < \sqrt{3}/38$, Proposition 6 shows that anomalous 5/2 phyllotaxis is possible. Other models of phyllotaxis in the cylindrical lattice assume phyllotaxis 2/1 (e.g., Mitchison's 2/1-phyllotaxis principle: see Research activity 1 in Section 5.6) instead of giving conditions determining it.

Proposition 6

The function $\min d(0, n)$ of d, where $1/3 < d < 1/2$, r and T fixed, has only one maximum for $r \geq \sqrt{3}/38$ (arbitrary T) or for $T < 5$ (arbitrary r). At this maximum we have phyllotaxis 2/1 if $r > \sqrt{3}/14$, and phyllotaxis 3/2 if $r < \sqrt{3}/14$. For $T = 5$ and $r < \sqrt{3}/38$, it has two maxima, at which the phyllotaxis is either 5/3 or 5/2. For $T > 5$ and $r < \sqrt{3}/38$ it has two or more maxima, and all types of phyllotaxis are possible (Adler, 1977a).

Proof. The intersection of line $r = c$ $(r < \sqrt{3}/6 : d < 1/2)$ with arc (n_{i+1}, n_i) determines a value of d for which $\min d(0, n)$ is maximized. For

$T < 5$ and $r \geq \sqrt{3}/38$, only two of these arcs are relevant, and $r = c$ intersects one or two arcs. In the first case the given function is maximized on arc $(2, 1)$, and in the second it is maximized on the upper portion of arc $(2, 1)$ and on the lower portion of arc $(3, 2)$ [starting at the point of intersection of arc $(3, 2)$ with arc $(2, 1)$]. For $T = 5$ and $r < \sqrt{3}/38$, given the relations in regions I–IV (see Figure 2.9), the maximization occurs on arc $(5, 3)$ and on the portion of the circle which includes arc $(5, 2)$ and which encloses region III (we cannot speak of this arc for $d < 2/5$, since in that case 3 is the next point of close return after 2), that is, for two values of d. When $T > 5$ and $r < \sqrt{3}/38$, many more arcs become relevant so that there may be more than two maxima. ∎

Exercises

2.20 (a) Prove the following identities:

$$F(k)^2 - F(k-1)F(k+1) = (-1)^{k-1}, \qquad (2.14)$$

$$F(k)F(k+1) - F(k-1)F(k+2) = (-1)^{k-1}, \qquad (2.15)$$

$$S_{t,k}F(k-1) - S_{t,k-1}F(k) = (-1)^k. \qquad (2.16)$$

(b) Under the hypothesis of Proposition 2 in Section 2.4.1, if $n = F(k)$ and $m = F(k+1)$ show that

$$(m - n)D_n + (2n - m)D_m = d \text{ or } 1 - d. \qquad (2.17)$$

(c) In the context of part b, show that the divergence angle $d < \frac{1}{2}$ is inside the bounds

$$\frac{(m-n)}{m} \quad \text{and} \quad \frac{n}{(m+n)}. \qquad (2.18)$$

(d) If the girth of the cylinder is C, instead of 1, show that

$$mD_n + nD_m = C, \qquad (2.19)$$

$$(m - n)D_n + (2n - m)D_m = dC \text{ or } (1 - d)C. \qquad (2.20)$$

(e) Under the hypothesis that (m, n) is but a visible pair, show that the bounds obtained for d include those given in part c.

2.21 (a) Prove that if d is between $F(k)/S_{t,k}$ and $F(k-1)/S_{t,k-1}$, $k \geq 3$, then $D_i = 3D_{i-2} - D_{i-4}$ for $i = 5, 6, \ldots, k+1$, where D_i stands for D_{n_i}.

(b) Prove that if $F(2K)/S_{t,2k} < d < F(2k-1)/S_{t,2k-1}$, $k \geq 1$, then $F(2k-2i)/F(2k-2i+2) < D_{2i+1}/D_{2i-1} < F(2k-2i-1)/F(2k-2i+1)$, and $1/3 < D_{2i+1}/D_{2i-1} < 1/2$, for $i = 1, 2, \ldots, k-1$, where D_i stands for D_{n_i}.

(c) Prove that if $F(2k)/S_{t,2k} < d < F(2k+1)/S_{t,2k+1}$, $k \geq 1$, then $F(2k-2i)/F(2k-2i+2) < D_{2i+2}/D_{2i} < F(2k-2i-1)/F(2k-2i+1)$, and $1/3 < D_{2i+2}/D_{2i} < 1/2$ for $i = 1, 2, \ldots, k-1$, where D_i stands for D_{n_i}.

2.22 Given a distribution of leaves with a constant divergence angle d such that $1/(t+1) < d < 1/t$, and $1/3 < D_{h+1}/D_{h-1} < 1/2$, for $h = 2, \ldots, k$, show that $F(2j+2)/S_{t,2j+2} < d < F(2j+1)/S_{t,2j+1}$, where $j = h/2$ if h is even, and $(h-1)/2$ if h is odd.

2.23 Show that if $d = (t + \phi^{-1})^{-1}$, $n_i = S_{t,i-1}$ for $i \geq 1$, $\lim_{k \to \infty} D_{i+1}/D_{i-1} = \phi^{-2}$ for $i \geq 2$, $D_i = (\phi^{-1})^{i-1}d$ for $i \geq 1$; for the case $t = 2$, that is, $d = \phi^{-2}$, we have $D_i = \phi^{-i-1}$ for $i \geq 1$.

2.24 If $d = (t + \phi^{-1})^{-1}$, determine the conditions on the rise r such that a given pair of opposed parastichies is conspicuous.

2.25 Imagine a system of circular leaves having the same diameter K, and suppose that the leaves $n = F(k-1)$, $m = F(k)$, and $m + n = F(k+1)$ are neighbors of the origin, equidistant from the leaf at the origin, and that leaf $m + n$ touches leaves m, n, and 0. If (x_n, y_n), (x_m, y_m), and (x_{m+n}, y_{m+n}) are respectively the coordinates of n, m, and $m + n$, where $x_n < 0$, show that, for large k, $y_n \cong K\sqrt{3}/2\phi\sqrt{2}$, $x_n \cong -K\phi^2/2\sqrt{2}$, and $x_m \cong K\sqrt{5}/2\sqrt{2}$.

2.26 Show that K in Exercise 2.25 has the value

$$K = C(m^2 + n^2 + mn)^{-1/2}, \tag{2.21}$$

where C is the girth of the cylinder. Show that when the leaves n, m, and $m + n$ are replaced as neighbors of the origin by the leaves m, $m + n$, $2m + n$, C increases by the factor ϕ.

2.27 Consider the system whose phyllotaxis rises from $n/(m-n)$ to m/n to $(m+n)/m$, where $m = F(k+1)$ and $n = F(k)$. Show that the maximal value of D_m/D_n is $(m + 2n)/(n + 2m)$. Show that the divergence is within the bounds $(3n^2 - nm + m^2)/2(m^2 + n^2 + mn)$ and $(3m^2 - 7mn + 5n^2)/2(n^2 - mn + m^2)$ (obtained by Mitchison, 1977), themselves included in the bounds given in Expression 2.18. Conclude that the increase of k brings the divergence nearer to ϕ^{-2}.

2.28 If d is in the following intervals, determine consecutive points of close return and the relevant arcs (n_{i+1}, n_i) (assuming that $T \geq n_{i+1}$): (a) $1/3 < d < 1/2$; (b) $1/3 < d < 2/5$; (c) $2/5 < d < 1/2$; (d) $2/5 < d < 3/7$; (e) $3/8 < d < 2/5$; and (f) $1/6 < d < 1/5$.

2.29 In the (d, r) plane draw the arcs $(2, 1)$, $(3, 2)$, $(5, 2)$, $(5, 3)$ and determine the permitted values for d.

2.30 (a) If $3/8 \leq d \leq 2/5$ and $T \leq 5$, what is the maximal value of r for which leaf 5 is a neighbor of the origin? Calculate the corresponding value of d.

   (b) If $2/5 \leq d \leq 3/7$ and $T \leq 5$, what is the maximal value of r for which leaf 5 is a neighbor of the origin?

   (c) If $1/3 \leq d \leq 1/2$ and $T \leq 5$, what is the maximal value of r for which leaf 5 can be a neighbor of the origin? Answer the same question for leaf 2, $T \leq 2$.

2.31 What will Expression 2.10 become if the girth of the cylinder is 2π?

2.32 Show that when (d, r) is on the arc $(F(k), F(k - 1))$, δ_k, the maximal diameter of the leaves, is given by

$$\delta_k^2 = (-1)^k \frac{2F(k)F(k - 1)d - (-1)^k - 2F(k)F(k - 2)}{F(k - 2)F(k + 1)}. \quad (2.22)$$

2.33 Let $1/(t + 1) < d < 1/t$, $t \geq 2$, and let (d_k, r_k) be the point of intersection of arc $(S_{t, k-1}, S_{t, k-2})$ with arc $(S_{t, k}, S_{t, k-1})$. Show that

$$d_k = \frac{4F(k - 1)S_{t, k-1} + (2t - 1)(-1)^{k-1}}{2\left[2S_{t, k-1}^2 + (-1)^{k-1}(t^2 - t - 1)\right]}, \quad (2.23)$$

$$r_k = \frac{\sqrt{3}}{2\left[2S_{t, k-1}^2 + (-1)^{k-1}(t^2 - t - 1)\right]} \quad (2.24)$$

(from Adler, 1974). Show that the closed nested intervals $[d_k, d_{k-1}]$

or $[d_{k-1}, d_k]$, $k > 1$, determine the point $(t + \phi^{-1})^{-1}$. If the point (d, r) is always confined to these arcs and $t = 2$, that is, if the system shows normal rising phyllotaxis with $t = 2$, draw the phyllotactic biography. This is the *PATH OF NORMAL FIBONACCI PHYLLOTAXIS*. Show that for $t = 2$, the interval above coincides with the one in Exercise 2.27.

2.34 If the point (d, r) moves on the arc $(21, 58)$, r decreasing until three leaves are equidistant from the origin, show that the third leaf is 79, and that $15/79 \le d \le 4/21$. If (d, r) moves on arc (m, n) as above, what is the number of the third leaf?

2.5 RESEARCH ACTIVITIES

1. *Complement on continued fractions.*

   (a) Prove the laws of formation of the convergents of $\omega = [a_0; a_1, a_2, a_3, \ldots] > 0$, given in Expression 1.3, with the initial values $p_0 = a_0$, $q_0 = 1$, $p_1 = a_0 a_1 + 1$, $q_1 = a_1$ (it is sometimes useful to put $p_{-1} = 1$, $q_{-1} = 0$, $p_{-2} = 0$, and $q_{-2} = 1$, and to consider the relations from $k \ge 0$). The proof in Khintchine (1963) is the most simple.

   (b) Show that p_n and q_n are relatively prime (see Exercise 1.7).

   (c) Putting $c_n = p_n/q_n$, $n = 0, 1, 2, \ldots$, prove that

$$c_{n-1} - c_n = \frac{(-1)^n}{q_n q_{n-1}}, \qquad n \ge 1, \tag{2.25}$$

$$c_{n-2} - c_n = \frac{a_n (-1)^{n-1}}{q_n q_{n-2}}, \qquad n \ge 2. \tag{2.26}$$

   (d) The odd convergents c_{2n+1} of an infinite continued fraction form a decreasing sequence, and the even convergents c_{2n} form an increasing sequence. Show that each c_n, $n \ge 2$, is between c_{n-1} and c_{n-2}.

   (e) Prove that the sequence $c_0, c_1, c_2, c_3, \ldots$ converges toward a limit greater than any even convergent and smaller than any odd convergent.

(f) Prove that the sequence $\langle c_n \rangle$, $n = 1, 2, 3, \ldots$ converges toward ω. Conclude that ω is greater than any even convergent and smaller than any odd convergent.

(g) Let $\omega_{n+1} = [a_{n+1}; a_{n+2}, \ldots]$. Show that $\omega = (\omega_{n+1}p_n + p_{n-1})/(\omega_{n+1}q_n + q_{n-1})$, and that

$$|\omega - c_n| < |\omega - c_{n-1}| \tag{2.27}$$

[for Expression 2.27, show that $(\omega - c_n) = (\omega - c_{n-1})(-q_{n-1}/\omega_{n+1}q_n)$; see Khintchine, 1963].

(h) Show that in Klein's diagram for the continued fraction of ω, the straight line passing through the points $(0, 0)$ and (q_n, p_n) tends to coincide with the line $y = \omega x$ as n increases, alternately from either side of this line (compare with the theorem in Section 1.4.1).

(i) Show that consecutive points of close return coincide with the denominators of consecutive principal convergents of the continued fraction of the divergence d, and alternate on either side of the vertical axis of the cylindrical lattice. Notice that by Exercises 1.12 and 1.23, if q_{n-1} is a point of close return, the next point of close return is necessarily on the other side of the vertical axis and is one of the following points: $q_{n,1}, q_{n,2}, \ldots, q_{n, a_n - 1}, q_n$. By Exercise 1.27, and by part g above, we have $|q_{n-1}d - p_{n-1}| > |q_n d - p_n|$. But it must be shown that $|q_{n-1}d - p_{n-1}| < |q_{n,c}d - p_{n,c}|$, for $c = 1, 2, \ldots, a_n - 1$, $n = 1, 2, 3 \ldots$. It would follow that q_n is the next point of close return after q_{n-1}. Khintchine (1963, pp. 27–35) has an interesting approach to this result. He presents the notions of better approximations of order 1 of a number (and shows that they are the principal and intermediate convergents) and of order 2 (and shows that they are the principal convergents). Leveque (1956) proves this result with Farey sequences.

2. *Audiovisual in phyllotaxis.* The phenomenon of phyllotaxis and the mathematical concepts extracted from it lend themselves to the audiovisual treatment. Here are a few suggestions of material to prepare.

(a) With scissors and thin cardboard make the cylinderlike lattices proposed by Hallé (1979b). What use can be made of these?

(b) Present McCulloch's method for building three-dimensional cones (see Kilmer, 1971; Jean, 1978b, pp. 169–170).

(c) Explain Church's method (1904, p. 53) for constructing the capitulum of the sunflower (see Jean, 1978b, Appendix I).

(d) Prepare material for the overhead projector to support an elementary report on the subject of phyllotaxis.

(e) Plan a color film on the phenomenon of phyllotaxis or an animated film on the rudiments of botanometry.

(f) See also Research Activities 8 of Chapter 2 and 1, 4, and 6 of Chapter 3.

3. *Photosynthesis and phyllotaxis.* Starting from the idea that photosynthesis requires a minimal superposition of the leaves on a stem, Leigh (1972) proposes a mathematical argument aiming to prove that the divergence angle ϕ^{-2} gives the leaves a maximal exposure to the sun, standing still at the zenith. Make a detailed presentation of his utilization of continued fractions. What exactly does he prove? (See Section 5 of the Epilogue, "The Functional Problem".)

4. *Patterns in microorganisms.* Fine biological structures present patterns that can be compared to those observed in phyllotaxis. Frey-Wyssling (1954) reports divergence angles of 5/18, 8/29, and 13/47 between individual amino acid residues in α-polypeptide synthetic chains (poly-*l*-alanine, poly-γ-methyl-*l*-glutamate), corresponding to helix periods of 27 Å, 43 Å, or 70.4 Å. He noticed that these fractions belong to the sequence $\frac{1}{3}, \frac{1}{4}, \frac{2}{7}, \frac{3}{11}, \frac{5}{18}, \frac{8}{29}, \frac{13}{47}, \frac{21}{76}, \ldots$, made with the consecutive convergents of the continued fraction of $(3 + \phi^{-1})^{-1}$, corresponding to normal phyllotaxis with $t = 3$ (see Exercise 1.8). Figure 1.8 represents the cylindrical lattice of these chains. Completely unaware of the language of phyllotaxis, Abdulnur and Laki (1983) have used this angle to distribute 142 residues of α-helices (tropomyosin, haemagglutinin) on 7-parastichies in the centric representation (Chapter 3). The 4- and 11-parastichies, though not traced, are conspicuous.

According to Erickson (1973) the classical terminology of phyllotaxis is quite suitable to describe the microscopic arrangements of the monomers on the microtubules of many organisms, such as the flagella of bacteria, the microfilaments of actin, and the protein coats of viruses. He deduced, from van Iterson's mathematical data (1907), and as far as the quality of the electron micrographs allowed it, the phyllotaxis or symmetry of these microorganisms. It is known that van Iterson calculated, for small spheres gathered on the surface of a cylinder and presenting the arrangement denoted by $J(n, m, m + n)$, $J = 1, 2, 3, \ldots$, parameters such as the divergence angle, the secondary divergence, the rise r, the radius of the cylinder, and the slopes of the parastichies. Plants also sometimes present J leaves on the same level of the stem; the system is then described as *MULTIJUGATE*. One can see J equidistant "genetic spirals" and J identical helicoidal systems, as in the case of van Iterson's spheres. When $J = 1$, the symbolism $(n, m, m + n)$ simply indicates that the phyllotaxis of the system is, at the same time, m/n and $(m + n)/m$.

The photograph on the left in Figure 2.10 is a bacteriophage showing the arrangement 4(2, 3, 5) of its hexamers. Actin has (1, 2) symmetry, the flagella of *Salmonella* presents the pattern 2(2, 3, 5), tobacco mosaic virus has phyllotaxis (1, 16, 17), turnip yellow mosaic virus suggests the pattern (17, 26, 43), and many microtubules have the pattern (6, 7, 13). These numbers belong to a normal phyllotaxis sequence for different values of t. The number $m^2 + mn + n^2$ found in Exercise 2.26 appears in the analysis of the triangle-faced polyhedra, models of the symmetry of polyhedral viruses; the number K of the same exercise is the diameter of the monomers on the microtubule showing phyllotaxis m/n.

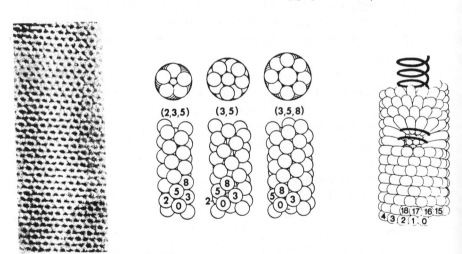

Figure 2.10   Studies of patterns of spheres on cylinders (by van Iterson, 1907), and application to the analysis of the symmetry of microorganisms, such as the one on the left (from Erickson, 1973). The illustrations in the middle show the rising of the phyllotaxis from 3/2 to 8/5. On the right, we have a model of the protein coat of the Tobacco mosaic virus. (Copyright 1973 by the American Association for the Advancement of Science.)

Harris and Erickson (1980) present a geometric analysis of the morphology of these tubular arrays of spheres. Make a detailed presentation of it, establishing relations with the concepts of Chapter 2, as already outlined by Adler (1977b) (see Frey-Wyssling, 1975).

5. *Various types of phyllotaxis.*

(a) Beside the cases of alternate phyllotaxis, normal and anomalous, with their respective sequences of integers and divergence angles, there are patterns described as decussate (frequent in dicotyledons), distichous (frequent in monocotyledons), bijugate, multijugate, spirodistichous, semitricussate, orthodistichous, pseudobijugate, subdecussate, tetrastichous, and spiromonostichous. One can even observe different types of phyllotaxis or the passage, natural or induced, from one type to another on a given plant (see Research Activities 5 in Section 5.6, and 7 in Section 3.4). Describe and illustrate the types named and their mutual relations, with examples of plants exhibiting them. Consult Church (1968), Roberts (1977), Richards and Schwabe (1969), Fujita (1937, 1942), Williams (1975), Rutishauser (1981), Schoute (1936), Dormer (1972), Wardlaw (1965a, 1968a), Cutter (1966), and Loiseau (1969). Are there as many clockwise as counterclockwise genetic spirals? (See Wardlaw, 1965a; Sinnott, 1960; Davis, 1964; Loiseau, 1969; Majumdar, 1948.)

(b) Present the proof given by Bravais and Bravais (1837, pp. 47, 54) of the following result: if (m, n) is a visible opposed parastichy pair, the genetic spiral is unique if and only if m and n are relatively prime numbers. Conclude that the number of leaves in a whorl (bunch of equidistant leaves on the same level of the stem: $J > 1$ in Research Activity 4), is the greatest common divisor of m and n.

(c) We already know (part a) that there is a variety of phyllotactic patterns. But these can be classified under three headings: *ALTERNATE, DISTICHOUS*, and *WHORLED*. According to Richards the alternate, *DECUSSATE*, and whorled systems are essentially the same, having one, two, three, or many primordia on the same level of the stem. The alternate system has two helices winding in opposite directions, each one going through all the primordia. The distichous and decussate systems are limiting cases of spiral unijugate or bijugate (multijugate with $J = 1$ and 2) systems, respectively. In a distichous system the leaves are distributed on two opposed vertical columns, one leaf at each level. The decussate system presents a pair of leaves at each level, each pair forming a right angle with those two pairs on the adjacent levels of the stem. In a bijugate system, two leaves arise at each plastochrone, and the divergence angle is 68.75°, that is, half of 137.51°, and two genetic spirals can be drawn. When many leaves appear on the same level of a stem, more or less simultaneously, we have the so-called whorled phyllotaxis. The systems of alternate or superposed whorls (see Jean 1978b, p. 102) belong to the class of *MULTIJUGATE SYSTEMS*, where the numbers of contact parastichies are generally multiples of the numbers in the normal phyllotaxis sequence (Expression 2.1). These systems embrace all the others. Formulate a few of the results of Chapter 2 for the multijugate systems (see Adler, 1974, and Section 5.2.2).

6. *The phenomenon of ramification.* Starting with the idea that buds are potentially capable of unlimited growth, Franquin (1974) proposes a model of spatiotemporal growth of a stem, systematically developing all the buds of all its boughs. Outline the manner in which the main series arises in the process (see Jean, 1978b, pp. 166–168).

7. *Swierczkowski's theorem.* Marzec and Kappraff (1983) consider the set Φ_G of angles whose continued fractions do not have intermediate convergents from a term forward (this set contains the angles investigated by Coxeter; see Section 1.4.3). They show that these angles distribute the leaves around the stem of a plant more uniformly than the neighboring angles can do. From the results in Research Activity 1, prove Swierczkowski's and Knuth's theorem, that is, Steinhaus' conjecture, show that the angles in Φ_G possess the property presented in this theorem, and that these angles guarantee the preceding spacing property of the leaves, as formalized by the first two authors.

8. *Graphic determination of the possible visible opposed parastichy pairs.* Draw a spiral in the plane, with a small pitch, and distribute 25–30 points on it, separated by ϕ^{-2} or 137°30′29″. Then erase the spiral, and make several copies of the set of points. Now (1) draw all possible families of regularly spaced spirals through the points by the shortest path between them around the common pole, one family per copy, and (2) superpose the points of two different copies so as to obtain visible

opposed spiral pairs. Notice that the possible families of spirals obtained in step 1 are those with $1, 2, 3, 5, 7, 8, 11, 12, 13, 14, 17, \ldots$ spirals, but that step 2 produces only the pairs $(1, 2), (2, 3), (3, 5), (5, 8), (8, 13), \ldots$ (see the graphs of Stevens, 1978, pp. 158, 159, 161). Check the results by working graphically with the m-parastichies in the cylindrical lattice displayed in Figure 2.1. Explain intuitively why the other families of spirals $(4, 6, 9, 10, 15, 16, \ldots)$ are rejected. Notice that Adler's theorem of Section 2.2.3 allows us to determine *all* visible pairs by means of the first algorithm of Section 2.3.2. Show that an opposed pair (m, n) is visible if and only if $mD_n + nD_m = 1$, D_n and D_m being the absolute values of the secondary divergences of n and m, respectively, in the normalized cylindrical lattice.

3 THE CENTRIC REPRESENTATION OF PHYLLOTAXIS

A description of a problem is a particular approach eventually leading to an original solution.

3.1 PRESENTATION

As for a crystal symmetry, it is possible to define the regular and repetitive arrangement of the primordia gathered around an apex by means of just a few numerical constants. The study of botanometry implies two distinct problems, namely, the description of any system in terms allowing its complete reconstruction and the formulation of theories explaining its origin. This chapter concerns the first problem, as it has been dealt with by Richards and presented by Jean (1979b). The parameters used in Richards' description are the divergence angle, the plastochrone ratio, and the phyllotaxis index in the plane, together with the necessary adjustments on the approximately conical surface of the apex. Our aim is to expose the essential relations between these parameters and to illustrate them by means of examples drawn from Church (1904), Thomas (1975), and Thornley (1975b). Our sources are the works of 1948, 1951, and 1956 of Richards, and his posthumous work presented by Schwabe in 1969. But the exposition set forth by Richards is not always sufficiently detailed and precise to be followed easily. Even in recent publications on the subject, the authors may mystify the reader by the unexplained use of double logarithms and bizarre formulas, in an ambiguous method of presentation punctuated by important omissions. A systematic presentation is carried out here in the form of a chain of propositions, or micro-graduated problems, requiring only elementary analytical geometry. Sometimes hints are given to facilitate the demonstrations further.

It is hoped that this chapter will reduce the misery of botanists generated by their efforts to master what is considered an article of faith (see Research Activity 6 in Section 3.4). We shall be able to appreciate the precision of the views of Richards, the pioneer whose papers contain other relevant theoretical and experimental considerations (see Research Activity 3). Richards, a botanist, favored an explanation of phyllotaxis that would be based on a chemical field theory, where each primordium exerts an inhibiting influence on the growth of others, an influence that decreases with the distance and that is ruled by Fick's laws of diffusion (presented in Chapter 4). He never developed field equations, however. This has been done only recently, by many authors, as we see in Chapter 5. We see also, in the Epilogue, why the diffusion theory of an inhibitor is insufficient to explain the angle ϕ^{-2}.

It is well-known that the organization and functioning of the plant apex is profoundly influenced by differential growth. According to Wardlaw (1965a), "The position and size of the leaf growth-centres are fundamentally determined by the genetically-controlled allometric growth-pattern." Allometry is the simplest expression of differential growth; it springs from the constant ratio of the relative growth rates of two variables. Section 3.3.5 presents allometric relations (from Jean, 1983g) between the radial primordial spacing represented by the rise r, or the plastochrone ratio R, and the visible pair (m, n). In his famous book D'Arcy Thompson (1942) complained that the developments in the domain of phyllotaxis were limited to the study of the so-called divergence angles of the systems. Though this parameter is fundamental in the description of phyllotaxis, and is in the background of the developments in Section 3.3.5, it expresses only the tangential primordial spacing. In many ways the radial spacing is of more immediate interest and usefulness.

3.2 PARAMETERS OF THE CENTRIC REPRESENTATION

3.2.1 A Bit of History

In the classical method, called the *SPIRAL THEORY OF SCHIMPER AND BRAUN*, just one constant was considered enough to describe the phyllotactic systems, namely, the divergence. It could only be a rational number, that is, one of the terms of Schimper and Braun's series (Chapter 1). At the end of the nineteenth century, Sachs (1882) and Hofmeister (1868) rejected this theory of the 1830s, and Church (1904) transferred the debate to the idea of contact parastichies radiating around a central pole (see Figure 1.1), as for the capitulum of a sunflower or a transverse section, assumed to be circular, of an apex. The opposed parastichies are, by definition, families of logarithmic spirals; in each family, each spiral can be reached from any other one by consecutive rotations of a constant angle around the common pole. Thus the word "centric" is used to qualify this plane representation. As for the cylindrical representation, an opposed parastichy pair can be visible and conspicuous.

CHURCH'S EQUIPOTENTIAL THEORY (1904, 1920) assumed that the opposed parastichies are orthogonal at the foliar insertions; that is, the tangents to the opposed curves at their points of intersection make right angles. These spirals completely characterized the patterns. This theory did not receive the experimental support it needed; the universality of orthogonal intersection has not been confirmed, and the surgical experiments of Snow and Snow, in the 1930s, have nearly demolished it.

To each phyllotactic system, Church associated the *BULK RATIO*, defined, in the case of a primordium assumed to be circular, as the ratio of its radius to the distance from its center to the center of the transverse section of the apex, which is the sine of half the angle at the center subtended by the primordium. This parameter represents one of the first attempts to define the *RADIAL SPACING* of the primordia, as opposed to the *TANGENTIAL SPACING* determined by the *DIVERGENCE*. Schwendener and Schoute have used more or less equivalent parameters, now replaced by Richards' *PLASTOCHRONE RATIO* (1948). Van Iterson (1907) used the reciprocal of this ratio under the name of relative radius.

Then came *PLANTEFOL'S MULTIPLE FOLIAR HELICES THEORY* (1948, 1950), where the interest centered on one of the two families of Church's contact parastichies. Camefort (1956) presents a complete report of it. In spite of notable successes, this theory, which avoids the mathematical treatment, failed to solve the problem of explaining the origin of the primordial patterns.

From this quest at least remain the notions of orthostichy, contact parastichy, and plastochrone ratio, elements of the vocabulary currently used in the centric representation of phyllotaxis. This vocabulary has recently been enriched (see Research Activity 2), and formulas have been developed to pass to the cylindrical representation analyzed in the preceding chapters (see Research Activity 5).

3.2.2   Richards' Approach

RICHARDS' DESCRIPTIVE CENTRIC THEORY is meant to be independent of any assumption about the origin of the primordial patterns. According to Richards (1951), for a complete description three parameters are necessary: the divergence angle, the plastochrone ratio, and the angle at the vertex of the cone tangent to the apex in the zone under consideration. In the case of a transverse section, the first two are sufficient to localize the primordia. A function of the plastochrone ratio, called the phyllotaxis index, gives immediately, by means of curves (Section 3.3.2), information that allows us to compare the various systems, thus making of phyllotaxis a continuous function. This index is related to the ratio of the surface of the central apex to that of the new primordium, that is, the radial relative growth rate of the apex (Section 3.3.5, and Research Activity 3).

The *DIVERGENCE ANGLE* (*d*) between two consecutively born primordia is the angle at the center determined by the radii linking their centers to the pole of the centric representation. The *PLASTOCHRONE* (*T*), usually measured in days, is the time elapsed between the births of these two primordia. The *PLASTOCHRONE RATIO* (*R*), a number greater than 1, is the ratio of the distances of their centers to the pole. The *PHYLLOTAXIS INDEX* (P.I.) is a quantity obtained from *R*, and defined by

$$\text{P.I.} = 0.379 - 2.3925 \log \log R. \tag{3.1}$$

The *EQUIVALENT PLASTOCHRONE RATIO* is the quantity

$$R_c = R^{\csc \alpha}, \tag{3.2}$$

where 2α is the angle at the vertex of the cone tangent to the apex in the zone under consideration. The *EQUIVALENT PHYLLOTAXIS INDEX* (E.P.I.) is defined by

$$\text{E.P.I} = \text{P.I.} + 2.3925 \log \sin \alpha. \tag{3.3}$$

Sections 3.3.2 and 3.3.3 make explicit the meaning of Expressions 3.1–3.3. The natural logarithm is denoted by ln, and the logarithm to base 10 by log. As in the preceding chapters, *d* is a number between 0 and $\frac{1}{2}$, and the corresponding angle in radians is denoted by δ.

3.2.3   On the Logarithmic Spiral in Phyllotaxis

Beyond the vague attraction exerted by the decorative and mystical properties of the spirals, culminating in Goethe's spiral theory of the 1830s, the botanical investigations of spiral growth forms began with Bonnet in the middle of the eighteenth century, and continued with the brothers Bravais in the middle of the nineteenth century. We have seen that the latter authors represented the phyllotactic patterns in terms of helices twisting round a cylinder.

Initially, the phyllotactic relations were expressed by means of the spiral of Archimedes. This is a plane curve whose polar equation is $r = a\theta$ (see

Research Activity 4). Observation reveals that the logarithmic spiral is more suitable for the description of the relative arrangements of the primordia on plants. Let us imagine an empirical spiral drawn through consecutively born primordia located in a plane, or orthogonally projected onto a plane. Let us suppose that the divergence is constant, as is also the time T in plastochrones. The angular rate of growth is then constant: $d\theta/dT = k_1$. In the tissues of young plants, it is generally found that the linear dimensions grow at a constant relative growth rate $dr/r\,dT = k_2$. Now let

$$r = ke^{\theta \cot \phi}, \quad \text{or} \quad \ln r = k + \theta \cot \phi \tag{3.4}$$

be a logarithmic spiral, in polar coordinates (r, θ), where r and θ are functions of T, and k and ϕ are constants. We have

$$\frac{dr}{r\,dT} = \frac{\cot \phi\, d\theta}{dT},$$

and the relative growth rate of r is constant. In a system that undergoes uniform development, the introduction of the logarithmic spiral is thus quite justified. The logarithmic nature of the parastichies rests on two assumptions, namely the uniformity of the tangential primordial spacing and the constancy of the plastochrone ratio.

Problem 3.1. Determine, in polar coordinates, the equation of the curve such that the tangent at its point (r, θ) makes a constant angle ϕ with the vector passing through this point.

If (r_1, θ_1) and (r_2, θ_2) are two points on this curve, show that

$$\frac{r_1}{r_2} = e^{(\theta_1 - \theta_2)\cot \phi}. \tag{3.5}$$

Problem 3.2. Let (m, n) be a visible opposed parastichy pair, in the centric representation of a leaf distribution with divergence d and plastochrone ratio R. Suppose that the opposed parastichies are logarithmic spirals with constant angles ϕ_m and ϕ_n, respectively. Let ψ_m and ψ_n, respectively, be the angles made by the vectors passing through two consecutive leaves on a spiral. Show that

$$m \ln R = \psi_m \cot \phi_m, \tag{3.6}$$

$$n \ln R = \psi_n \cot \phi_n, \tag{3.7}$$

$$\frac{\cot \phi_m}{\cot \phi_n} = \frac{m\psi_n}{n\psi_m}, \tag{3.8}$$

$$\psi_m = 2\pi p - 2\pi md, \tag{3.9}$$

$$\psi_n = 2\pi q - 2\pi nd, \tag{3.10}$$

where p and q (the encyclic numbers of m and n) are the integers nearest to md and nd, respectively, ψ_m and ψ_n being of opposite signs and between $-\pi$ and π.

Finally, show that the angle of intersection β between two opposed parastichies is (> 0 by convention) given by

$$\beta = \phi_n - \phi_m, \quad \text{and} \quad mn \ln^2 R - 2\pi \ln R \cot \beta + \psi_n \psi_m = 0 \tag{3.11}$$

(compare with Expression 2.5). Figure 3.1 illustrates this problem.

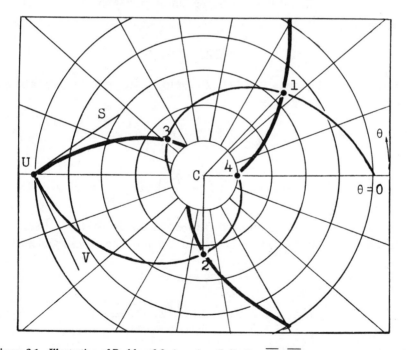

Figure 3.1 Illustration of Problem 3.2: $(m, n) = (3, 2)$; $R = \overline{C1}/\overline{C2} = 1.5$; $d = $ angle $2C1 = 3/8$; $\psi_2 = $ angle $4C2 = -\pi/2$; $\psi_3 = $ angle $4C1 = \pi/4$; $\phi_2 = $ angle $CUV = -63°$; $\phi_3 = $ angle $CUS = 33°$; $\beta = 96°$.

3.2.4 Fundamental Properties

Problem 3.3. Let (m, n) be a visible opposed parastichy pair of a distribution of leaves with divergence d. Let us suppose that the youngest primordium, the nearest to the pole, has $(1, 0)$ as polar coordinates. Show that the equations of the n-parastichies, and m-parastichies are respectively given by

$$r_n = \exp\left(\theta_n - \frac{2\pi j}{n}\right)\cot\phi_n, \quad j = 0, 1, \ldots, n - 1,$$

$$r_m = \exp\left(\theta_m - \frac{2\pi i}{m}\right)\cot\phi_m, \quad i = 0, 1, \ldots, m - 1,$$

(3.12)

where for $j = 0$ and $i = 0$, the spirals meet at the youngest leaf.

If q_m (respectively q_n) is the ratio of the lengths of the vectors passing through two consecutive leaves along an m-parastichy (respectively n-parastichy), prove that

$$q_m^n = q_n^m, \quad m\psi_n - n\psi_m = \pm 2\pi. \tag{3.13}$$

Show that at the leaves P and Q, nearest to the leaf at $(1, 0)$, on $j = 0$ and $i = 0$, respectively, we have

$$\psi_n\cot\phi_n = \left(\psi_n \mp \frac{2\pi}{m}\right)\cot\phi_m, \tag{3.14}$$

$$\psi_m\cot\phi_m = \left(\psi_m \pm \frac{2\pi}{n}\right)\cot\phi_n, \tag{3.15}$$

where the $+$ sign in Expression 3.14 corresponds to the $-$ sign in Expression 3.15, and vice versa. Show that if p and q are as in Problem 3.2,

$$np - mq = \pm 1, \tag{3.16}$$

$$d = \frac{1}{2}\left(\frac{q}{n} + \frac{p}{m}\right) \pm \frac{1}{2}\frac{(\cot\phi_n + \cot\phi_m)}{mn(\cot\phi_n - \cot\phi_m)}, \tag{3.17}$$

$$\ln R = \pm \frac{2\pi \cot\phi_n\cot\phi_m}{mn(\cot\phi_m - \cot\phi_n)}. \tag{3.18}$$

Hint. At point P, $j = 0$, $i = \pm 1$, $\theta_n = \theta_m = \psi_n$, and $r_n = r_m$. To find ψ_n and ψ_m, one can also use the transformation $x = \ln r$, $y = \theta$. To obtain Expressions 3.16 and 3.17, substitute ψ_n and ψ_m from Expressions 3.14 and 3.15 in Expressions 3.9 and 3.10. For $\ln R$, use Expression 3.14 or 3.15 with Expression 3.6 or 3.7.

Problem 3.4. Let (m, n) be a visible opposed parastichy pair of a distribution of leaves with divergence d. If m and n are relatively prime, show that $np - mq = \pm 1$ (Expression 3.16) has at most two solutions.

We have met Expression 3.16 many times in the context of the preceding chapters (Expressions 1.11 and 2.4). Indeterminate analysis teaches that this equation can have integral solutions only if m and n are relatively prime. Since ψ_m and ψ_n have opposite signs, the solution, if it exists, is unique. In other respects, forgetting the restrictions on p and q, this equation possesses an infinity of solutions. It is obvious that if d, R, m, and n are fixed, one can choose any solution to determine $\psi_m, \psi_n, \phi_n, \phi_m$ from the preceding equations. But then ψ_m and ψ_n are not necessarily opposed in sign, and the pair (m, n) may not be opposed though it will be "visible," but not necessarily "conspicuous." Thornley (1975b, pp. 516–517) illustrated this situation and deduced the value of the divergence given in Expression 3.17. Richards (1951) showed that for d and R fixed, many systems of opposed parastichies are possible, depending on the shape of the primordia. It is also possible to draw many pairs (m, n) for d, R, m, and n fixed. But the visibility of opposed families is more than a visual effect; it is, as we have seen in Chapter 2, an indication of the value of the divergence, and a pair is discernible to the eye, that is conspicuous, if ψ_m and ψ_n are opposite in sign and very small. The following result is more general than the one illustrated by Thornley for the main series.

Problem 3.5. Under the conditions stated in Problem 3.4, show that if

$$m = S_{t, k+1} \quad \text{and} \quad n = S_{t, k},$$
(3.19)

then either

$$q = F(k) \quad \text{and} \quad p = F(k + 1),$$
(3.20)

or

$$q = F(k)t - F(k - 2) \quad \text{and} \quad p = F(k + 1)t - F(k - 1). \qquad (3.21)$$

Expression 3.21 can be obtained by subtracting Expressions 3.20 and 3.19: $(n, m) - (q, p) = (n - q, m - p)$, and if $mq - np = \pm 1$, $m(n - q) - n(m - p) = \mp 1$.

Problem 3.6. State a result similar to the one in Problem 3.5 for the sequence $\langle R_{t, k} \rangle$ defined by $R_{t, k} = F(k + 1)t + F(k - 1)$, $k = 1, 2, 3, \ldots, t \geq 2$.

Hint. Determine the consecutive convergents of the continued fraction of $(t + \phi^{-2})^{-1}$.

Problem 3.7. Under the hypothesis of Problem 3.3, show that if ψ_m and ψ_n are small enough

$$d \simeq \frac{q/n + p/m}{2}, \qquad (3.22)$$

or

$$d \simeq \frac{(n - q)/n + (m - p)/m}{2}, \qquad (3.23)$$

and verify that the sum of these two expressions is 1. With m, n, p, and q as in Expressions 3.19 and 3.20, show that d in Expression 3.22 is always less than or equal to $\frac{1}{2}$ for $t \geq 2$, and greater than or equal to $\frac{1}{2}$ for $t = 1$. Conclude that if ψ_n and ψ_m are equal to zero, the divergence is rational, and the leaves are distributed on orthostichies. If d has the rational value given in Expression 3.22, and m and n the values given in Expression 3.19, verify that $|\psi_m/\psi_n| \simeq \phi$.

Hint. Use Expressions 3.6, 3.7, and 3.17.

One can compare the terms in the arithmetic mean of Expression 3.22 with the end points of the interval given in the theorem of Section 1.3.2, for a divergence $d \leq \frac{1}{2}$ and $t \geq 2$ (when $d \leq \frac{1}{2}$, $mq - np = \pm 1$ alternately, with the increase of k).

Problem 3.8. Let (m, n) be a visible opposed parastichy pair of a distribution of leaves with divergence $d = \phi^{-2}$. If $m = F(k + 1)$ and $n = F(k)$, show that

$$\psi_m = 2\pi(-1)^k \phi^{-k-1}, \tag{3.24}$$

$$\psi_n = 2\pi(-1)^{k-1} \phi^{-k}, \tag{3.25}$$

$$\frac{\psi_n}{\psi_m} = -\phi. \tag{3.26}$$

Hint. Use the identity $F(k)\phi - F(k + 1) = (-1)^{k+1}\phi^{-k}$.

Problem 3.9. Let (m, n) be a visible opposed parastichy pair of a distribution of leaves with divergence $d = (t + \phi^{-1})^{-1}$. Prove that $\psi_n/\psi_m = -\phi$.

3.3 SELECTED TOPICS

3.3.1 The Case of Orthogonal Parastichies

Problem 3.10. If the opposed parastichies of the pair (m, n) are orthogonal, show that

$$\ln R = \left(\frac{-\psi_m \psi_n}{mn} \right)^{1/2}, \tag{3.27}$$

$$\cot^2 \phi_m = \frac{-m\psi_n}{n\psi_m}, \quad \cot^2 \phi_n = \frac{-n\psi_m}{m\psi_n}. \tag{3.28}$$

Hint. Use Expressions 3.6, 3.8, and 3.11.

Figure 3.2 illustrates Expression 3.27. Given (m, n), then p, q (Expression 3.16), ψ_m, and ψ_n (Expressions 3.9 and 3.10) are fixed, and $\ln R$ becomes a function of d. The angles being in degrees, the right side of Expression 3.27 must be multiplied by $\pi/180$.

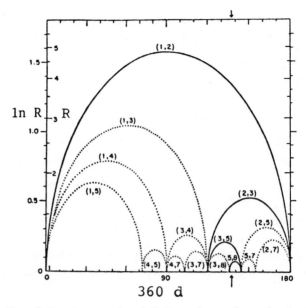

Figure 3.2 The relations between d and R for various orthogonal pairs (m, n) (after Maksymowych and Erickson, 1977).

Problem 3.11. Let (m, n) be a visible opposed parastichy pair of a distribution of leaves with divergence $d \leq 1/2$, where m and n are as in Expression 3.19, and

$$d = \frac{q + p}{m + n},$$ (3.29)

where p and q are given in Expression 3.20. Show that

$$|\psi_m| = |\psi_n| = \frac{2\pi}{m + n}, \qquad \left|\frac{\psi_m}{\psi_n}\right| = 1.$$ (3.30)

$$\left|\frac{\cot \phi_m}{\cot \phi_n}\right| = \frac{m}{n},$$ (3.31)

$$\cot \phi_m = \frac{-(n + m)\cot \beta \pm \left[(n + m)^2 \cot^2 \beta + 4mn\right]^{1/2}}{2n}$$ (3.32)

where β is given in Expression 3.11. Deduce that if the opposed parastichies are orthogonal,

$$\ln R = \frac{|\psi_m|}{(mn)^{1/2}} = \frac{|\psi_n|}{(mn)^{1/2}}. \qquad (3.33)$$

$$\cot^2 \phi_m = \frac{m}{n}, \qquad \cot^2 \phi_n = \frac{n}{m}. \qquad (3.34)$$

Thomas (1975) obtained the results in Problem 3.11, except for Expression 3.32, for the particular case of the main series. Expression 3.29 has not been explicitly formulated by Thomas. I deduced it from his values for ψ_m and ψ_n (Expression 3.30) by means of Expressions 3.9, 3.10, and 3.16, and from the consideration of his various figures. This value e for the divergence is the mediant between the end points of the interval given in the theorem of Section 1.3.2. If d is in the open interval $(q/n, e)$, or $(e, p/m)$ (if $q/n < p/m$), then leaf $m + n$ is the next close return candidate after m. This means that Thomas' orthostichies can emerge only at the precise instant when the phyllotaxis of a system changes to a phyllotaxis with larger secondary numbers.

Problem 3.12. Let the system of m and n opposed spirals be defined by

$$n \ln r = n \ln c + m\theta + (2k - 1)\pi, \qquad (3.35)$$

$$m \ln r = m \ln c - n\theta + (2l - 1)\pi, \qquad (3.36)$$

(Church's equations, 1904) in a centric representation (r, θ) of the primordia where m and n are relatively prime. Show that

$$\psi_m = \pm \frac{2\pi n}{m^2 + n^2}, \qquad (3.37)$$

$$\psi_n = \mp \frac{2\pi m}{m^2 + n^2}, \qquad (3.38)$$

$$\ln R = \frac{2\pi}{m^2 + n^2}. \qquad (3.39)$$

Show that Church's bulk ratio (Section 3.2.1) is given by the formula

$$\sin\left(\frac{\pi}{(m^2 + n^2)^{1/2}} \right). \qquad (3.40)$$

If m and n, $m > n$, are two consecutive terms of the main series, show that the divergence is given by

$$d = \frac{(m - n)^2 + n^2}{m^2 + n^2}.$$ (3.41)

Verify that $np - mq = \pm 1$.

Hint. Use the orthogonal transformation $x = \ln r$ and $y = \theta$ to obtain two families of orthogonal straight lines. Show that these lines determine a tessellation of squares whose sides are equal to $2\pi(m^2 + n^2)^{1/2}$. Consider the spirals $l = 0$, $l = 1$, $k = 0$, $k = 1$; the center of the square defined by the corresponding straight lines is $(\ln c, 0)$, and the equation of the inscribed circle becomes, in the initial system, $[\ln(r/c)]^2 + \theta^2 = \pi^2/(m^2 + n^2)$. Notice that

$$\cot \phi_m = \pm \frac{m}{n}, \qquad \cot \phi_n = \mp \frac{n}{m},$$ (3.42)

(reciprocal and orthogonal spirals), and that $k = 0, 1, \ldots, m - 1$, and $l = 0, 1, \ldots, n - 1$. The upper (respectively lower) signs for ψ_n, ψ_m, $\cot \phi_m$, $\cot \phi_n$ go together. Church (1920, p. 340) obtained $d^* = (2mn - m^2)/(m^2 + n^2)$, instead of Expression 3.41. In fact, $d^* = 1 - d$ when m and n are inverted.

The rational divergences in Expressions 3.29 and 3.41 are determined along orthostichies. In the cases of Schimper and Braun and of Thomas, there are $m + n$ orthostichies; in the case of Church, there are $m^2 + n^2$. In the last case, the value of d is nearer to ϕ^{-2}, ψ_n/ψ_m tends toward $-\phi$, and if m and n are very large, Expression 3.22 gives a good approximation for this value of d. In other respects, in the shoot apex of alternate systems, where the leaves are initiated, no superposed primordia, corresponding to a rational divergence, can be observed. The irrational divergence $(t + \phi^{-1})^{-1}$ "is no longer a mystical conception of aim on the part of the plant, but, within the range of one degree, an actual mathematical property of Fibonacci phyllotaxis for all ratios 3/2 and upwards" (Church, 1920, p. 18). When the divergence of a specimen is determined by means of Schimper and Braun's foliar cycle, or by Bravais' formula, the point ultimately chosen as making an orthostichy is only an approximation. The systems in Problems 3.11 and 3.12 are thus very particular.

Problem 3.13. Under the conditions stated in Problem 3.8, if the opposed parastichies are orthogonal, show that

$$\log R = 2\pi\phi^{-1}\log e\left(\phi^{2k-1}mn\right)^{-1/2}, \qquad (3.43)$$

$$\log R = 1.68646192\left(0.618033989^{2k-1}/mn\right)^{1/2} \qquad (3.44)$$

(Expression 3.44 is found in Richards, 1951, p. 518).

Problem 3.14. Given a system with phyllotaxis $m/n = F(k+1)/F(k)$ and divergence $d = \phi^{-2}$, show that

$$R = e^{2\pi r} \qquad (3.45)$$

where r is the rise in the normalized cylindrical representation. Show that the result can be extended to $m/n = S_{t,k}/S_{t,k-1}$, corresponding to the sequence in Expression 2.1 and to the sequence in Expression 2.2.

Hint. Consider Expressions 3.24, 3.25, and 3.27 and then Coxeter's formula in Chapter 1. The normal phyllotaxis sequence can be dealt with Expression 1.16.

3.3.2 Richards' Phyllotaxis Index

Problem 3.15. Let (m, n) be a visible opposed parastichy pair for a distribution of leaves with divergence $d = \phi^{-2}$, and angle of intersection β. If $m = F(k+1)$ and $n = F(k)$, show that

$$\log R = (-1)^{k-1}\pi\phi^{-k-1}$$
$$\times\left[(n + m\phi)\cot\beta \pm\left((n + m\phi)^2\cot^2\beta + 4mn\phi\right)^{1/2}\right]\log e/mn.$$

$$(3.46)$$

Verify that $\beta = \pi/2$ gives Expressions 3.27 and 3.43.

Figure 3.3 shows for $d = \phi^{-2}$ that given R, β of a pair (m, n) can be estimated, and that given β of a pair (m, n), R can be estimated. Moreover, if the value of P.I., calculated from R, is an integer, then the figure allows us to conclude that the conspicuous opposed parastichies are exactly orthogonal. Richards built tables to obtain the phyllotaxis index easily, given R and the angle of intersection β of the parastichy pairs. There are also tables giving R from β.

Problem 3.16.   Let (m, n) and $(m + n, m)$ be two consecutive orthogonal visible opposed parastichy pairs, where m and n are consecutive terms of the main series and $d = \phi^{-2}$. If R_1 and R_2 are the respective plastochrone ratios, prove that for large m

$$\frac{1}{\log \phi^2} (\log\log R_1 - \log\log R_2) \simeq 1, \tag{3.47}$$

$$\text{P.I.} \simeq N, \tag{3.48}$$

where N is an integer.

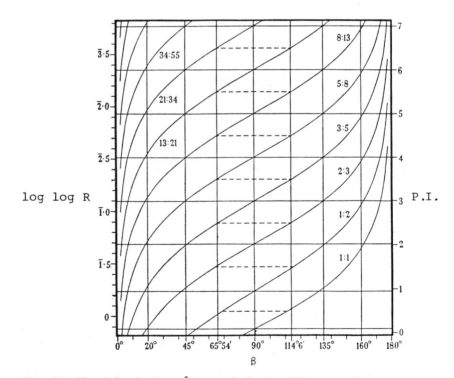

Figure 3.3   The relation, for $d = \phi^{-2}$, between $\log\log R$ and β for consecutive pairs (m, n). In particular, if $\beta = 90°$, P.I. is approximately an integer for all pairs (m, n) (from Richards, 1951).

It was to simplify the use of R that Richards (1948) introduced the phyllotaxis index. In his definition (Expression 3.1), the constant 2.3925 is an approximation of $1/\log \phi^2$ found in Problem 3.16, and the constant 0.379 is what must be added to the expression $1 - 2.3925 \log\log R_1$ (calculated for m and n large) to obtain an integer. The phyllotaxis index is thus defined in such a way that consecutive integers correspond to consecutive orthogonal visible opposed parastichy pairs when $d = \phi^{-2}$.

Problem 3.17. Let (m, n) be an orthogonal visible opposed parastichy pair, of a distribution of leaves with divergence d. Show that, as a function of d, the phyllotaxis index is a curve with two asymptotes, symmetrical with regard to a vertical axis.

Problem 3.18. Let $(m, n) = (3, 2)$ be a visible opposed parastichy pair of a distribution of leaves with divergence $\delta = 2\pi d$. Show that

$$\ln R = \frac{\pi \cot \beta \pm \left(\pi^2 \cot^2 \beta - 24\pi^2 + 60\pi\delta - 36\delta^2 \right)^{1/2}}{6}. \quad (3.49)$$

For $\beta = \pi/2$ determine the equations of the asymptotes ($\delta = \pi$ and $\delta = 2\pi/3$), and of the axis of symmetry ($\delta = 5\pi/6$) mentioned in Problem 3.17.

Figure 3.4 gives the curves for some pairs (m, n) and some values of β (denoted $m : n, \beta$ in the figure), of the phyllotaxis index as a function of d. The letter F stands for the angle ϕ^{-2}. All these curves have the same form, their lower portion being rather flat. The curves drawn in heavy lines cut $d = F$ at an integral P.I., and the larger portions of consecutive curves oscillate on either side of this straight line. Two points have been chosen by Richards on each curve, and between them the phyllotaxis index differs by 0.13 only from an integer; there is thus a good range of values for d, around $d = F$, for which the phyllotaxis index deviates by a very small amount from an integer. Richards has illustrated, for the pair $(2, 3)$, the curves corresponding to different values of β. Problems 3.17 and 3.18 give supplementary informations on these curves. Expressions 3.27, 3.43, and 3.49 coincide when $m = 3$, $n = 2$, $\beta = \pi/2$, and $\delta = 2\pi\phi^{-2}$. Figure 3.4 can be compared to Figure 3.2.

Problem 3.19. Let (m, n) be an orthogonal visible opposed parastichy pair of a distribution of leaves with divergence $d = \phi^{-2}$. If $m = F(k + 1)$, and

$n = F(k)$, show that

$$\text{P.I.} = \frac{k}{2} + 1.19625 \log mn - 0.414 \tag{3.50}$$

(this formula is found in Richards, 1951, p. 534).

"Measures are needed whose validity and usefulness do not depend on the assumptions of some particular phyllotaxis theory" (Richards, 1951, p. 513). But the phyllotaxis index seems to lean heavily on the assumption of the orthogonality of spirals, which forced Richards to reject Church's equipotential theory (Section 3.2.1). But this index is nothing but a point of reference, among many, to compare the systems. Richards even calculated

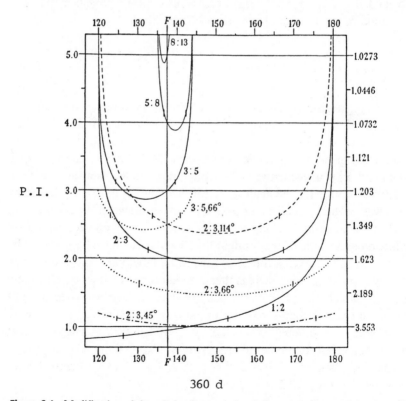

Figure 3.4 Modification of the relation between the phyllotaxis index and β by the variation of d (from Richards, 1951).

this index when $d = (t + \phi^{-1})^{-1}$ (1951, p. 562); obviously he does not any longer obtain integers for orthogonal pairs. Thomas (1975) modified Richards' index so as to obtain approximately an integer when the parastichies are orthogonal and d is his rational divergence; he then "corrects" Richards and states that the phyllotaxis index is equal to $0.465316 - 2.392486 \log \log R$. This index differs from Richards' P.I. by $0.05 < 0.13$. In short, one can calculate such indexes for a wide range of arbitrary divergences. Richards' system of references is the less arbitrary.

3.3.3 The Equivalent Phyllotaxis Index

The centric representation is perfectly adapted to the study of the primordia as they appear on a flat apex, such as for the sunflower or the Compositae in general. A transverse section of a bud, for example, does not give an exact representation of the relations between the parastichies on the surface of this bud. In order to describe the system of curves, as observed on this surface, the plastochrone ratio R must be adjusted to obtain the equivalent plastochrone ratio R_c. A typical apical surface has the form of a paraboloid, locally equivalent to the surface of a cone where the youngest primordia are inserted. Looking at Figure 3.5, let us consider the parameters related to primordia i and $i + 1$. Let $R'_{i+1} = \rho_i/\rho_{i+1}$ be their plastochrone ratio in the zone of cone $\rho'_i = \rho'_{i+1} + w$, θ_{i+1} the angle of the sector of ring subtending the arc $\rho_i\theta_{i+1}$, and d_{i+1} and d'_{i+1} the fractions of a turn corresponding to the angles θ_{i+1} and θ'_{i+1}. Let us stretch tangentially the sector of ring, so as to bring the divergence θ_{i+1} on a complete ring in the plane (it can be calculated that the stretching factor is equal to $\csc \alpha_i$). In order to obtain a centric representation and to preserve the angles, a radial stretching of the sector of ring must also be performed (factor $\csc \alpha_i$). Problem 3.20 shows that the new plastochrone ratio $R_{i+1} = \rho''_i/\rho'_{i+1}$ is related to R'_{i+1} by Expression 3.2.

Problem 3.20. Prove Richards' transformation (1951):

$$\ln R_{i+1} \simeq \csc \alpha_i \ln R'_{i+1}.$$

Hint. Show that $R'_{i+1} = \rho'_i/\rho'_{i+1}$, $\theta'_{i+1} = \rho_i\theta_{i+1}/\rho'_i$, $d'_{i+1} = d_{i+1} \sin \alpha_i$, the angle of the sector of ring is $2\pi \sin \alpha_i$, and the stretching of the sector of

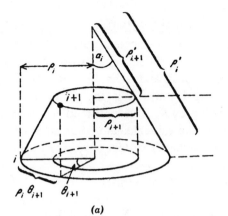

(a)

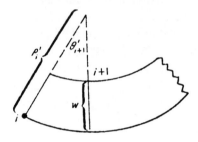

(b)

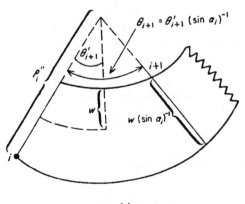

(c)

Figure 3.5 Transformation of a zone of a conical surface into a zone of a disk. (a) Zone of the conical surface. (b) Development of the zone in the plane: sector of ring before stretching. (c) After radial and tangential stretching of the sector. [From Adler, 1977a; with permission from *Journal of Theoretical Biology*, **65**, 1977. Copyright: Academic Press Inc. (London) Ltd.]

ring replaces ρ_i' by $\rho_i'' = \rho_{i+1}' + w\,\csc\,\alpha_i$, $R_{i+1} - 1 = (R_{i+1}' - 1)\csc\,\alpha_i$. Since R_{i+1} and R_{i+1}' are numbers of the form $1 + x$ for small x, one can complete the argument with the help of the formula $\ln|1 + x| = x - x^2/2 + x^3/3 - x^4/4 + \cdots, -1 < x \leq 1$.

Since Expression 3.47 is approximately equal to 1 when R_1 and R_2 are replaced by the corresponding equivalent plastochrone ratios, it follows that

$$\text{E.P.I.} = 0.379 - 2.3925 \log\log R_c,$$

$$\text{E.P.I.} = \text{P.I.} + 2.3925 \log\sin\alpha. \tag{3.3}$$

E.P.I. is related to R_c as P.I. is to R.

Problem 3.21. Prove Richards' formula (1956, p. 68):

$$\text{E.P.I.} = 1.246 + 2.3925 \log\!\left(\frac{P}{w}\right) \tag{3.51}$$

where

$$P = \frac{\rho_i - \rho_{i+1}}{\ln\rho_i - \ln\rho_{i+1}}.$$

Hint.

$$\log R_c = 0.43429\,\csc\,\alpha\,\ln R,$$

$$\log\log R_c = -0.3622 - \log\!\left(\frac{P}{w}\right).$$

3.3.4 van Iterson's Formula

Problem 3.22. Prove van Iterson's formula for the visible system (m, n) of circular tangent primordia, in the centric representation, where $\delta = 2\pi d$:

$$\frac{\cos(m\delta/2)}{\cos(n\delta/2)} = \pm\frac{1 + R^m}{1 + R^n}R^{(n-m)/2}. \tag{3.52}$$

Expression 3.52 cannot be solved directly. Following van Iterson (1907, 1960), Maksymowych and Erickson (1977) produced, with the help of the computer, the curve of the relation between R and d for a variety of pairs (m, n) (Figure 3.6). This figure can be compared to Figures 2.8 and 2.9, built in the cylindrical representation; d is a function, respectively, of ln R, of $d(0, m)$, and of r.

The model in Problem 3.22 can be obtained by a conformal transformation of the cylindrical lattice of Chapter 2. The tangent circles with equal diameters of the cylindrical representation become the tangent circles of Problem 3.22, the centers of the former moving on the contact circle of leaves m and n (see Section 5.6, Research Activity 1), more precisely on arc (m, n) (Section 2.4.2). It can be verified, for example, that the systems $(2, 3)$ and $(3, 5)$ are possible, in the case of van Iterson's tangent circles, for $128.17° \leq \delta \leq 142.11°$ and $135.42° \leq \delta \leq 142.11°$, respectively. But by Exercise 2.33, with $t = 2$, we have $d_2 = 5/14$, $d_3 = 15/38$, and $d_4 = 37/98$, that is, the preceding angles, and the abscissas of the points of intersection of the arcs $(2, 1)$ and $(3, 2)$, of the arcs $(3, 2)$ and $(5, 3)$, and of the arcs $(5, 3)$ and $(8, 5)$, respectively. It is ascertained once more that the cylindrical lattice allows us to deal with the problems of phyllotaxis with ease and precision; the relations between the parameters are not uselessly

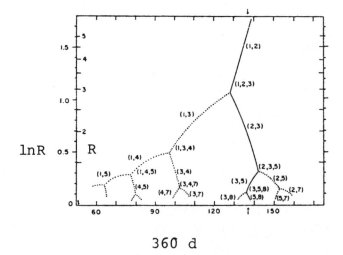

$$360 \; d$$

Figure 3.6 Relations between d and R for various pairs (m, n), ruled by van Iterson's formula for contiguous circles in the centric representation (from Maksymowych and Erickson, 1977).

complicated by transcendent functions, as in Expressions 2.9 and 3.52, on the cylinder and in the centric plane, respectively. Notice that in Figure 3.2, for the case of orthogonal systems, the values of d are very different from the preceding; that is, $1/3 \leq d \leq 1/2$ for the system $(2, 3)$ and $1/3 \leq d \leq 2/5$ for the system $(3, 5)$. The limits for d sometimes differ slightly, depending on the representation.

Van Iterson also proposed a model where the tangent circles, that is, the primordia, are replaced by the closed curves obtained from the projection in the plane of the contiguous circles on a cone whose angle at its vertex is equal to 2α. He obtains a formula similar to Expression 3.52 where md and nd are replaced by $\psi_m \sin \alpha$, and $\psi_n \sin \alpha$, respectively. Maksymowych and Erickson produced, with the help of a computer, graphs showing such systems of primordia.

3.3.5 Jean's Allometric Relation

Suppose that the *RELATIVE GROWTH RATES* of two variables X and Y, given respectively by $dX/X \, dT$ and $dY/Y \, dT$, experience throughout development a constant ratio, that is,

$$\frac{dY}{Y \, dT} = b \frac{dX}{X \, dT},\qquad (3.53)$$

a formula known as the differential form of the *ALLOMETRIC LAW*. Integration yields the usual form of this law of growth, that is,

$$Y = aX^b \qquad (3.54)$$

where a and b are constants, the latter being called the rate of allometry between the variables. It means that on double logarithmic paper the relation between X and Y is a straight line with slope b. This formula is closely linked to the logistic formulas of absolute growth representing S-shaped curves, such as Gompertz' function, and to Thompson's famous transformation theory (see Jean, 1983e).

As underlined by Wardlaw (1965a, p. 246) "For a given divergence angle d, the contact parastichies (m, n) depend on the plastochrone ratio R and the primordium shape." This section precisely points to an interdependence between the conspicuous parastichy pair (m, n) determining the phyllotaxis of the system, the plastochrone ratio R, and the angle of intersection β of the opposed parastichies. In order to establish this dependence, we must not speak about the shape of the primordia and the contact parastichies; though intuitively correct, these confusing ideas are replaced by those of visible and conspicuous opposed parastichy pairs and their angles of intersection, as defined earlier.

From Expression 1.16 for orthogonal systems, it can be shown that

$$Y \cong 1.89443 X^{-2} \tag{3.55}$$

in the cylindrical representation, where $Y = r$ and $X = m + n$, (m, n) being the phyllotaxis of the system. Making use of Expression 3.45 we have

$$Y \cong 5.16942 X^{-2} \tag{3.56}$$

in the centric representation, where $Y = \log R$.

More generally, using Expressions 2.3–2.5, it can be shown (see Jean, 1983f) for large k, for any pair $(m(k), n(k))$ of visible opposed parastichies with angle of intersection β, and for the divergence angle corresponding to the sequence $\ldots, n, m, m + n, 2m + n, 3m + 2n, \ldots$ (see Adler's theorem in Section 2.3.1, or Exercise 2.14), that

$$r \cong p(\beta)(m + n)^{-2}, \tag{3.57}$$

where

$$p(\beta) = \frac{\phi^3 \left[\sqrt{5} \cot \beta + \left(5 \cot^2 \beta + 4 \right)^{1/2} \right]}{2\sqrt{5}}. \tag{3.58}$$

To derive these expressions we needed to calculate secondary divergences. In the cases of normal and anomalous phyllotaxis [the sequences $S_{t,k} = F(k)t + F(k - 1)$, and $R_{t,k} = 2F(k - 1)t + F(k) + F(k - 2)$ in Expressions 2.1 and 2.2, respectively] the secondary divergences are given, respectively, in Exercise 1.26, and by $(-1)^{k-2}/\phi^{k-2}(2\phi t + \phi + 2)$.

For small values of k, it can be calculated that the percentage of deviation between the exact value of r, as obtained from Expression 2.5, and the approximation formula given in Expression 3.57 is very small, generally much below 1% (see Jean, 1983g).

It can thus be concluded that for any given β, the allometric line

$$r = p(\beta)(m + n)^{-2} \qquad (3.59)$$

contains all types of phyllotaxis (m, n), including those visible pairs with low secondary numbers. For a given β of the visible pair (m, n) we can read r on log log paper from a line of slope -2, and intercept $p(\beta)$. Given r, and the visible pair, β can be estimated.

To use the best correlation available, especially when m and n are small, the conspicuous parastichy pair must be chosen. It can be shown indeed that for $\beta \in (0°, 180°)$, the percentage of deviation of r from the right side of Expression 3.57 has an absolute minimum around 90°. Thus if one observes a plant with phyllotaxis (m, n), plastochrone ratio R, and angle of intersection β of the parastichies, and if the divergence angle is the limiting angle derived from the series $..., n, m, m + n, 2m + n, ...$, then the point $(\log R, m + n)$ is on the correlation line of slope -2 and intercept $2\pi p(\beta) \log e$. By means of the relation $R = e^{2\pi r}$ (Expression 3.45, which can be easily extended), the allometric relations become indeed

$$\log R = 2\pi p(\beta) \log e (m + n)^{-2}. \qquad (3.60)$$

However, notice that on an actual shoot apex, considered as a surface of revolution, the expression $e^{2\pi r}$ is rather the equivalent plastochrone ratio R_c given in Expression 3.2 and in Problem 3.20, and R is the plastochrone ratio in a transverse section of the apex, so that Expression 3.60 must be multiplied by $\sin \alpha$, where α is as in Expression 3.2.

We already knew relations between R, (m, n), and β, such as Expression 3.46. The corresponding expressions for the other types of phyllotaxis and divergences are still more complex. Yet Expression 3.60 is very simple. Expressions such as 3.44 and 3.50 regarding the phyllotaxis index, obtained by Richards from the centric representation of phyllotaxis, can be deduced from the above general setting in the cylindrical representation. Many other special formulas, and relations, can be derived from these considerations. Tables such as Table 3.1 should be useful to botanists. It says in particular that if R and (m, n) are such that Expression 3.56 is verified, then it can be concluded that the conspicuous opposed parastichies are orthogonal.

Richards (1951) points out that R may be expressed entirely in terms of rates of change within the apex. It is often quoted from him that when the factors affecting phyllotaxis are constant, $\ln R/T$, where T is the plastochrone period, is equal to the *RADIAL RELATIVE GROWTH RATE* c of the whole apex, in the region of initiation. It is concluded that for any given divergence angle between the primordia, the phyllotaxis (m, n) of a system may be regarded as the resultant of the relative velocities of two growth processes, rate of expansion of the apex, and rate of production of primordia. It is known that the larger the primordia relative to the size of the apex, the closer they are to the center of the apical meristem, and the lower the phyllotaxis (m, n), that is, m and n are small integers. It is also observed that phyllotaxis is expressed by higher members of the Fibonacci-type sequence $\ldots, n, m, m + n, 2m + n, \ldots$ when the area of the apical meristem is larger compared to the area of the primordia; then R is relatively smaller. We have shown indeed, in this section, that for any given divergence angle and for any given angle of intersection β of the conspicuous parastichy pair (m, n), the relative growth rate of $\ln R$ is proportional to the relative growth rate of $m + n$, the constant of

Table 3.1   Illustrating the allometric relations between R and (m, n) (Expression 3.60) for various values of β (in degrees).

$m + n$	β								
	74	76	78	80	82	84	86	88	90
3	6.128	5.707	5.332	4.997	4.695	4.424	4.179	3.956	3.753
4	2.772	2.664	2.564	2.472	2.387	2.308	2.235	2.167	2.104
5	1.921	1.872	1.827	1.785	1.745	1.708	1.673	1.641	1.610
6	1.573	1.546	1.520	1.495	1.472	1.450	1.430	1.410	1.392
7	1.396	1.377	1.360	1.344	1.329	1.314	1.300	1.287	1.275
8	1.290	1.278	1.266	1.254	1.243	1.233	1.223	1.213	1.204
9	1.223	1.214	1.204	1.196	1.187	1.180	1.172	1.165	1.158
10	1.177	1.170	1.163	1.156	1.149	1.143	1.137	1.132	1.126
11	1.144	1.138	1.133	1.127	1.122	1.117	1.112	1.108	1.103
12	1.120	1.115	1.110	1.106	1.101	1.097	1.093	1.090	1.086
13	1.101	1.097	1.093	1.089	1.086	1.082	1.079	1.076	1.073
14	1.087	1.083	1.080	1.077	1.074	1.071	1.068	1.065	1.063
15	1.075	1.072	1.069	1.066	1.064	1.061	1.059	1.057	1.054

proportionality being -2. It follows from Expression 3.60 that

$$T = \left(\frac{2\pi p(\beta)}{c} \right)(m + n)^{-2}, \qquad (3.61)$$

a relation in agreement with Richards' observation (1951, p. 547) that when phyllotaxis rises along the former Fibonacci-type sequence, T normally diminishes since c experiences no permanent effect.

3.4   RESEARCH ACTIVITIES

1. *Constructing the sunflower from the genetic spiral.* In a recent article, Ridley (1982a) uses continued fractions to prove that the value $d = \phi^{-2}$ gives the theoretically most efficient packing of the seeds in the sunflower, when these are distributed on the spiral $\sqrt{n}\, e^{2\pi i n d}$, $n = 1, 2, 3, \ldots$, devised by Vogel (1979). Make a study of these articles to present the essence of the reasoning (see also Dixon, 1981). Compare with Mathai's and Davis' method (Research Activity 4).

2. *A new centric description.* Analyze Thomas and Cannell's (1980) description of phyllotactic systems, with their new notions of generative angle and orthochrone ratio.

3. *Relation of phyllotaxis to rates of growth within the shoot apex.* Growth can be considered as "the set of quantitative modifications arising during development, and translated by an increase in the dimensions, without significant change in the qualitative properties" (Heller, 1978). The criteria that can be thought of are the measures of the geometric dimensions, of the mass, of the dry weight, or of the fresh weight, and so on. If a variable y is retained as a criterion, one can measure its growth Δy during the time Δt (see Research Activity 8 below), its average absolute growth rate $\Delta y/\Delta t$, its absolute growth rate dy/dt, its relative growth $\Delta y/y$, its mean relative growth rate $\Delta y/y\Delta t$, and its *RELATIVE GROWTH RATE* $d\ln y/dt = (dy/dt)/y = y'(t)/y(t)$. Present the relations established by Richards between the relative growth rates in the shoot apex, P.I., and E.P.I. (see Richards and Schwabe, 1969; Richards, 1951, 1956; Schwabe, 1971; and Cutter, 1964, 1965; Jean, 1983f, 1983g).

4. *Constructing the sunflower from the parastichies.* Exhibit Davis' method (Davis and Mathai, 1973) for constructing sunflowers (and palm trees: Davis, 1970, 1971), and Mathai's mathematical explanation (Mathai and Davis, 1974). What is Church's opposition (1901; 1904, pp. 64–65) to any method of construction based on Archimedes' spiral? Let us recall that if $\lambda(t) = (x(t), y(t))$, $t \in [a, b]$, is a curve such that $\lambda'(t)$ exists and is continuous, then λ has a finite length given by $L(\lambda) = \int_a^b [x'(t)^2 + y'(t)^2]^{1/2} dt$. Research Activity 2c of Chapter 2 gives a method of construction based on the parastichies.

5. *Formulas transforming the parabolic, conical, and centric representations into the cylindrical representation.* By means of an appropriate transformation, any primordial distribution on a disk, a cone, or a surface of revolution can become a cylindrical distribution. The questions of the passage from the disk to the cylinder (Problem 3.14), and from the conical surface to the disk (Problem 3.20) have already been considered. A cylindrical representation of the zone of a cone is thus obtained by combining the respective expressions: $r_{i+1} = (2\pi \sin \alpha_i)^{-1} \ln R'_{i+1}$. When $\alpha_i = \pi/2$ for every T, the surface of revolution becomes a disk and the last formula recalls Expression 3.45.

   Adler constructed a cylindrical representation of Richards' model (Figure 2.2), and of Mathai and Davis' model (Research Activity 4) when the consecutive primordia of the sunflower are on a spiral of Archimedes (in the centric representation). He obtained $r(T) = (2\pi)^{-1} \ln\{(1 + aT)/[1 + a(T + 1)]\}$ where $a = b/c$, c being the distance from the newly born primordium to the center of the centric representation, and b the increase of this distance at each plastochrone T (the primordium gets away from the center with growth). From a parabolic apical dome he deduced the expression $r(T) = (4\pi)^{-1}[1 + 4a(T + b)]^{1/2} \ln[(T + b)/(T + b - 1)]$, where b is the vertical distance from the tip of the dome to the horizontal plane going through the primordium initiated near the dome. Present the detail of these operations, as developed by Adler (1977a, pp. 52–58). Coxeter (1969) gives a family of transformations allowing us to go from the cone to the cylinder, a special case of which is Richards' transformation (Expression 3.2).

6. *Richards' theory for botanists.* Realizing that Richards' descriptive theory is not understood by the majority of botanists, Williams (1975, Chapter 3 and Appendix 3) tries to demystify it and to apply it. Make a

detailed report on his considerations and his method for constructing orthogonal spiral systems from the plastochrone ratio.

7. *Modifying phyllotaxis with phytohormones; Maksymowych and Erickson's Experimental Method.* It has been observed that morphogenetical substances do not seem, for the most part, to have any effect on the phyllotaxis of plant shoots (Williams, 1975). This demonstrates the stability of the phyllotactic patterns in general. Maksymowych and Erickson (1977) modified the phyllotaxis of *Xanthium* with gibberellic acid, Schwabe (1971) did the same with *Chrysanthemum* using triiodobenzoic acid, and Snow and Snow (1937) and Ball (1944) with *Epilobium hirsutum* and *Tropaeolum majus*, respectively, using auxin. Report on these experiments, emphasizing the nature of the observed changes, and on the quantitative analysis carried out by these authors. Consult Cutter (1966), Wardlaw (1965a), Loiseau (1969).

In particular let us emphasize here that Maksymowych and Erickson (1977) described a method that proved to be useful in the comparison of apical material to existing mathematical models of phyllotaxis. It has been used by Erickson and Meicenheimer (1977) in their study of photoperiod induced change in phyllotaxis in *Xanthium*, and by Meicenheimer (1979) in his study of the changing phyllotaxis of *Ranunculus*. Figure 3.7 illustrates the method, which can be summarized as follows. For at least three consecutive primordia P_1, P_2, and P_3, the divergence angle α and the plastochrone ratio are assumed to be constant. It follows that the former parameter can be estimated as the larger angle between the chords d_1 and d_2, and that the latter parameter has the value d_2/d_1. The values of d_1, d_2, and α are measured, and it follows that $\overline{P_2C}$ is found to be equal to $[d_1^2 d_2^2/(d_1^2 + d_2^2 - $

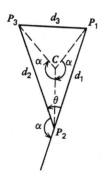

Figure 3.7 Geometric relationships involved in mensuration of divergence angles and radii from centers P_1, P_2, P_3 of primordia to center C of the apex, used in Maksymowych and Erickson's method.

$2d_1d_2\cos\alpha)]^{1/2}$. Illustrate the use of these estimates of the plastochrone ratio and the radii [see also Meicenheimer (1981)].

8. *Empirical models.* Beside the descriptive, mechanistic, and interpretative models dealt with in this book, empirical models, that is, fitted curves, occupy an important place in the treatment of plant growth. They are useful for summarizing the results and for making predictions from the data. Some processes appear to follow laws such as the exponential function $W = be^{kt}$, the logistic or autocatalytic function $W = a/(1 + be^{-kt})$, the monomolecular function $W = a(1 - be^{-kt})$, Gompertz' function $W = ae^{-be^{-kt}}$, and Richards' function $W = a(1 + be^{-kt})^{-1/n}$ which includes the previous functions as special cases. Make a presentation of these models of plant growth, introducing them from data gathered from plants, deducing their linear and differential forms and the relative growth rate of W (see Research Activity 3), and establishing relations with the law of allometry (Expression 3.54). See Richards (1959, 1969), Steward (1968), Williams (1975), Jean (1983e), Causton (1969, 1977), Hunt (1981), and Dormer (1972).

4 | THE DIFFUSION OF MORPHOGENS

One of the best examples of organized growth in plants is the regular arrangement of lateral organs, usually leaves, on a stem, and indeed the problems of phyllotaxis have received much attention; however they are still far from fully understood.
(Cutter, 1966)

4.1 PRESENTATION

When a colored substance, such as potassium permanganate, an iodized solution, or a drop of ink, is put into a container filled with water, a phenomenon of diffusion takes place: at first the substance is separated from water by a distinct boundary, then the molecules of the substance become evenly distributed in the water under the action of a concentration gradient. Where the concentration is high the number of molecules tends to decrease, and inversely, their number increases where the concentration is low. The molecules of the colored substance move under the impact of the molecules of water without any preference as to direction; it is a matter of pure chance that they become evenly distributed. This random process of migration, of transfer of matter inside a system, comes to an end when the density of the molecules is everywhere the same. Lehninger (1969) defines diffusion as the tendency of molecules to move in the direction of the smallest thermodynamic activity (concentration), so as to reach a uniform concentration, and a maximal entropy, in the system as a whole.

In 1913, Schoute, followed by Richards in 1948, postulated that the apical meristem, and each of the growing primordia, produces an inhibiting substance, whose diffusion prevents the formation of new primordia in their immediate neighborhood, where the concentration of the inhibitor is larger than a given threshold value. As soon as the concentration of the inhibitor falls below this value at a point, a new primordium is initiated at this point. In Figure 4.1, the arcs of circles represent the threshold of the concentration of the inhibitor produced by the primordia considered as points. A new primordium always arises at the lowest point of intersection of two or three arcs. This is the *DIFFUSION THEORY OF AN INHIBITOR*.

A famous experiment by Wardlaw (1949c) on the fern *Dryopteris aristata* seems to support this theory. He made two cuts on each side of a newborn primordium, and it developed more rapidly than usual. It is supposed that the primordium was thus released from the inhibiting influence of the neighboring primordia. However, Maksymowych and Erickson (1977) doubt this interpretation, contending that cells have been destroyed on both sides of the cuts.

Schoute has not been able to translate his theory into geometric terms. Recently, Thornley (1975a), Veen (1973), and Young (1978) proposed quantitative models of it, involving computer simulation. Meinhardt (1974) and Turing (1952) developed variants of this theory, where the activation

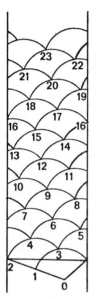

Figure 4.1   Phyllotaxis 2/1 and 3/2, produced by a mechanism of circular diffusion of an inhibitor secreted by point sources, the primordia, on a cylinder (from Schoute, 1913).

and the inhibition of chemical substances, reacting with one another, produce morphogenetic gradients.

The aim of this chapter is to prepare for the study of the models of diffusion analyzed in Chapter 5, that is, to familiarize ourselves with the process of diffusion and with the differential equations ruling it in the field of phyllotaxis. Section 4.2 is based on intuition, Section 4.3 is a mathematical deduction of Fick's law of diffusion from the general kinetic equation, and Section 4.4 is an analysis of the equations used by investigators in phyllotaxis. Broadly speaking, it is suggested that the reader concentrate first on Sections 4.2.1, 4.2.3, 4.4.1, and 4.4.2.

4.2   LAWS OF DIFFUSION

4.2.1   Fick's First Law, Concentration Gradient

The concentration, denoted by C, is a function of the time t and of the position r; that is, $C = C(r, t)$. It is supposed that $C(r, t)$ can be differentiated with regard to r and to t; the number of molecules involved is so large that the error made from the consideration of a regular function is unimportant. The quantity of matter diffusing through 1 cm^2/sec, in a given direction where the concentration is decreasing (minus sign in Expression 4.1), is expressed, in a system of Cartesian coordinates, by the formula

$$F = -D\frac{\partial C}{\partial x}. \qquad (4.1)$$

The factor $\partial C/\partial x$ is called the *CONCENTRATION GRADIENT*, F is the rate of transport per unit area of the section, in a direction x orthogonal to the section, and D is a factor of proportionality, the coefficient of diffusion, considered as a positive constant. Experimentation shows that this equation is rather precise when C is not too high.

Let us consider now a plane vertical membrane with area A and thickness Δx through which a substance diffuses. Let $M(x, t)$ be the mass of the migrating substance, on the left of the membrane at time t. On the left, and on the membrane, the concentration of the molecules is $C(x)$, and on the right it is $C(x + \Delta x)$, at time t. Experimental facts supporting the intuition allow us to state that the average rate of growth of the mass in the membrane, $\Delta M/\Delta t$, is proportional to $\Delta C/\Delta x$, the *AVERAGE CONCENTRATION*, and proportional to A. We thus have $\Delta M/\Delta t = DA\,\Delta C/\Delta x$. When $\Delta x \to 0$ we have $\Delta M/\Delta t = DA\,\partial C/\partial x$, and the mass flowing through the membrane at instant t is governed by the law

$$\frac{\partial M}{\partial t} = DA\frac{\partial C}{\partial x}. \qquad (4.2)$$

This phenomenological equation bears the name of Fick's first law of diffusion. It can be generalized to three dimensions. It is obvious that $M(x, t)$ grows with t if the flow of the molecules is toward the left. This happens if $C(x, t)$ increases with x, that is, if $\partial C/\partial x > 0$. If $\partial C/\partial x$ increases, it is intuitively clear that the rate of increase of M, that is, $\partial M/\partial t$, rises, and so does M.

4.2.2 Applications: Two Models of Diffusion

Let us consider, with Rashevsky (1961), a cell producing a metabolite, that is, an organic substance with low molecular weight participating in the reactions of the metabolism. Let us suppose that the permeability of the surface of the cell is so high that it does not offer any resistance to the flow of the metabolite, and that the concentration of the latter outside the cell is uniform, that is, C_0. Inside, near the surface, it is supposed that the concentration is also C_0. The metabolite in question is produced inside at an average rate of q g/cm³ · sec; there is an overflow that is expelled outside the cell. It is at the center 0 of the cell that the concentration is the highest. Let $\overline{C}$ be the average concentration inside the cell, midway between periphery and center. Let us calculate the rate of flow knowing the rate of production and the dimensions of the approximately cylindrical cell, illustrated in Figure 4.2.

The average change of the concentration along the vector **OB** is $\overline{C} - C_0$. The rate of variation in that direction is consequently $2(\overline{C} - C_0)/r_1$. It follows that the diffusion per square centimeter is $2D(\overline{C} - C_0)/r_1$. The total area of the section orthogonal to **OB** is approximately given by πr_2^2. Consequently, the total flow in the direction **OB** is $2\pi r_2^2 D(\overline{C} - C_0)/r_1$. The same quantity runs away in the direction **OB′** so that the total longitudinal flow is $4\pi r_2^2 D(\overline{C} - C_0)/r_1$.

There is also a radial flow in the direction **OA**. The rate of change of the concentration per square centimeter in that direction is $2D(\overline{C} - C_0)/r_2$. The total surface of passage is approximately equal to $4\pi r_1 r_2$ (cylinder whose height is $2r_1$). The total radial flow is thus $8\pi r_1 D(\overline{C} - C_0)$. It follows that the total rate of flow outside the cell, per second, of the metabolite is $4\pi D(2r_1^2 + r_2^2)(\overline{C} - C_0)/r_1$. At steady state, the quantity of metabolite flowing per second is equal to the quantity produced per second, that is, q times the volume of the cell. This volume being approximately equal to $2\pi r_2^2 r_1$, it follows that

$$2D(2r_1^2 + r_2^2)(\overline{C} - C_0) = r_2^2 r_1^2 q,$$

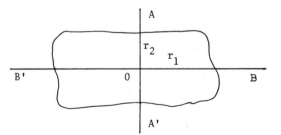

Figure 4.2 Calculation of the average concentration of a metabolite inside a cylindrical cell.

and finally

$$\overline{C} = C_0 + \frac{r_1^2 r_2^2}{2D(2r_1^2 + r_2^2)} q. \tag{4.3}$$

This expression gives the average concentration as a function of the dimensions of the cell, of the diffusion constant, of the rate of production, and of the constant external concentration. If q is proportional to $\overline{C}$, that is, if $q = \alpha\overline{C}$, $\overline{C}$ becomes proportional to C_0. Letting $r_1 = r_2$, we obtain an expression that is valid for an approximately spherical cell with radius r_1, that is,

$$\overline{C} = \frac{C_0}{1 + r_1^2 \alpha/6D}. \tag{4.4}$$

Let us consider now, with Thrall et al. (1967), a cell with constant volume suspended in a homogeneous liquid containing a dissolved substance whose constant concentration is $C_0 = C_0(x, t)$. Let $C(t)$ be the concentration of the dissolved substance inside the cell at time t, and let us suppose that this substance is distributed almost evenly inside the cell at any time, so that $C = C(t)$ depends only on time. By diffusion, molecules of the dissolved substance enter into the cell but also molecules of the substance go out of the cell. If $C_0 > C(t)$, the net flow is from the neighboring liquid toward the inside of the cell. Let us determine $C(t)$.

Let $M(t)$ be the mass of the dissolved substance in the cell, A the area of the cellular membrane, and V its volume. By definition of the concentration,

$$M(t) = VC(t). \tag{4.5}$$

Fick's law stipulates that the net flow rate, or rate of growth of M, that is, $dM/dt \,(= V\,dC/dt)$, is proportional to the area of the membrane and to the difference of concentration between the two sides of the membrane. We have $dM/dt = kA(C_0 - C)$. If $C < C_0$, M increases, and consequently $k > 0$. This constant is called the permeability of the membrane for the particular dissolved substance. We get $dC/dt = kA(C_0 - C)/V$, a separable differential equation for which

$$C = C_0 + ae^{-kAt/V} \tag{4.6}$$

where a is a constant of integration. When t goes to infinity, C asymptotically goes toward C_0.

4.2.3 Fick's Second Law: The Diffusion Equation

From Fick's first law we now derive an equation that proves to be more useful when solving certain problems. Let us return to the symbolism used when the first law was discussed, and let us consider that $M(x, t)$ is a function of x only. The volume occupied by the mass $\Delta M = M(x + \Delta x, t) - M(x, t)$ is equal to $A \Delta x$, so that the average density is $\Delta M / A \Delta x$. When $\Delta x \rightarrow 0$, the last expression tends to $C(x, t)$, so that

$$\frac{\partial M}{\partial x} = AC(x, t). \tag{4.7}$$

Given that by hypothesis all the second partial derivatives are continuous, we have $\partial^2 M / \partial x \, \partial t = \partial^2 M / \partial t \, \partial x$, and using Expressions 4.2 and 4.7 we have $\partial^2 M / \partial x \, \partial t = DA \, \partial^2 C / \partial x^2 = A \, \partial C / \partial t$, and

$$\frac{\partial C}{\partial t} = D \frac{\partial^2 C}{\partial x^2}. \tag{4.8}$$

This is Fick's second law, the famous one-dimensional diffusion equation. It can be generalized to three dimensions, that is, $\partial C / \partial t = D(\partial^2 C / \partial x^2 + \partial^2 C / \partial y^2 + \partial^2 C / \partial z^2)$, and more generally

$$\frac{\partial C}{\partial t} = D\nabla^2 C. \tag{4.9}$$

This law is identical to the law of heat conduction, the word concentration being replaced by the word temperature. It is precisely by exploiting the analogy between the two processes that Fick, a physiologist and biophysicist, obtained the diffusion equations in 1855; he essentially adapted the equations of heat developed by Fourier 30 years before. It is a simple law of an empirical nature, but deducible from the kinetic theory of solutions (see Smolukovski's equation in Section 4.3.2); however, the partial differential equation involved generally presents great difficulties. Its solution gives the concentration at each point, something that the biologist cannot measure in any way. Moreover, the law of diffusion is only an approximation; it is more valid if a large number of molecules is involved. On the other hand, the average concentration can be determined experimentally, as in Rashevsky's model of Section 4.2.2. A relatively coarse approximation of the law proves to be adequate in many cases.

Problem 4.1. Determine the coefficients a and b so that $C(x, t) = C^{ax+bt}$ satisfies Fick's diffusion equation.

Problem 4.2. Show that the diffusion equation in a cylinder and in a sphere (cylindrical and spherical coordinates) becomes, respectively,

$$\frac{\partial C}{\partial t} = D\left(\frac{\partial^2 C}{\partial \rho^2} + \frac{1}{\rho}\frac{\partial C}{\partial \rho} + \frac{1}{\rho^2}\frac{\partial^2 C}{\partial \theta^2} + \frac{\partial^2 C}{\partial z^2} \right), \qquad (4.10)$$

$$\frac{\partial C}{\partial t} = \frac{1}{\rho^2}\left[\frac{\partial}{\partial \rho}\left(D\rho^2 \frac{\partial C}{\partial \rho} \right) + \frac{1}{\sin\theta}\frac{\partial}{\partial \theta}\left(D\sin\theta \frac{\partial C}{\partial \theta} \right) + \frac{D}{\sin^2\theta}\frac{\partial^2 C}{\partial \phi^2} \right]. $$

$$(4.11)$$

For a radial diffusion only, verify that these equations become

$$\frac{\partial C}{\partial t} = \frac{1}{\rho}\frac{\partial}{\partial \rho}\left(\rho D \frac{\partial C}{\partial \rho} \right), \qquad (4.12)$$

$$\frac{\partial C}{\partial t} = D\left(\frac{\partial^2 C}{\partial \rho^2} + \frac{2}{\rho}\frac{\partial C}{\partial \rho} \right), \qquad (4.13)$$

and that for $\partial C/\partial t = 0$, we have $C = m + n\ln\rho$ in the first case, and $C = m + n/\rho$ in the second case.

4.3   ANALYSIS OF THE EQUATIONS

4.3.1   Steady-State Concentration

The analysis of the diffusion equation (Expression 4.8) reveals a few important properties. For example, if $C(x, t)$ is concave upward, $\partial^2 C/\partial x^2 > 0$, $\partial C/\partial t > 0$, and C is an increasing function of t, for all x. If C is a linear function of x, $\partial^2 C / \partial x^2 = 0$, $\partial C / \partial t = 0$, and as a function of t, C is a constant. This is the so-called steady-state concentration, not varying any more with time, at each point. The steady state, or state out of equilibrium of an open system, is that state where all the forces acting on a system are exactly compensated by opposite forces, so that all the components have a stationary concentration even though matter flows throughout the system (Lehninger, 1969).

The theories of diffusion in phyllotaxis generally suppose an inhibiting substance having a steady-state concentration. This means that

$$\frac{\partial C}{\partial t} = 0. \tag{4.14}$$

This arises because the process of growth of a plant is much slower than the process of diffusion; the plastochrone is measured in days whereas the time required for the diffusion of small molecules across an apex—that is, a few tenths of a millimeter—is a few minutes.

Let us consider a tube of length L and transverse section area A, linking two tubs where the concentrations of a morphogen are respectively given by C_0 and C_1. With Crick (1970), we can also imagine a row of cells whose total length is L, where the left cell produces a chemical substance, a morphogen, and maintains it at a constant concentration C_0, and the right cell maintains the concentration at C_1. After a certain time we have a steady-state system; that is, at every point of the tube, or of the row of cells, the concentration no longer varies. Since $\partial C / \partial t = 0$, $\partial^2 C / \partial x^2 = 0$, and it follows that $C = mx + n$. If $C = C_0$ at $x = 0$, and $C = C_1$ at $x = L$, as in Figure 4.3, we

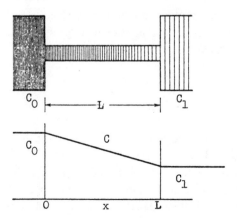

Figure 4.3 Straight line representing the steady-state concentration of a morphogen between two sources of constant concentrations, C_0, and C_1, respectively.

have

$$C = C_0 - \frac{(C_0 - C_1)x}{L},$$ (4.15)

the concentration gradient and the rate of diffusion being $-(C_0 - C_1)/L$ and $A(C_0 - C_1)/L$, respectively.

4.3.2 From Smolukovski's Equation to the Diffusion Equation

Following is the kinetic equation, also known as the differential form of Smolukovski's equation for Markov's processes:

$$\frac{\partial f(x, t/x_0, t_0)}{\partial t} = \sum_{n=1}^{\infty} \frac{(-1)^n}{n!} \frac{\partial^n}{\partial x^n} [A_n(x, t) f(x, t/x_0, t_0)].$$ (4.16)

The following special form is known as Fokker-Planck's equation:

$$\frac{\partial f}{\partial t} = -\frac{\partial}{\partial x} [A_1(x, t) f] + \frac{\partial^2}{2 \partial x^2} [A_2(x, t) f],$$ (4.17)

where f is a continuous probability density function and the A_i are the infinitesimal moments.

The kinetic equation is not very useful. However, the situation is improved when the moments vanish from a certain n. It has been proved that if the kinetic equation contains only a finite number of derivatives, it is at most an equation of order 2. Three partial differential equations of order 2 very often appear in the applications and dominate all the theory of partial differential equations. These equations are

$$\frac{\partial \mu}{\partial t} = \alpha^2 \frac{\partial^2 \mu}{\partial x^2}, \ \frac{\partial^2 \mu}{\partial t^2} = \beta^2 \frac{\partial^2 \mu}{\partial x^2}, \ \frac{\partial^2 \mu}{\partial x^2} + \frac{\partial^2 \mu}{\partial y^2} = 0.$$

We recognize the first one, known as the heat equation. The second one is the wave equation, and the third is Laplace's potential equation. These equations are generally accompanied by boundary conditions, dictated by the nature of the problem under consideration; they are chosen so as to guarantee a unique solution. Section 4.3.3 illustrates the so-called method of

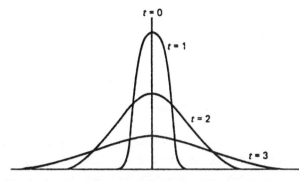

Figure 4.4 Density function f of a normal distribution with mean x_0 (the state of the system at t_0) and variance $\sigma^2(t - t_0)$ for various values of t.

separation of the variables to solve these equations. To that purpose here is a simple theorem from the theory of differential equations.

Consider the homogeneous linear equation of order n with constant coefficients

$$a_0\frac{d^n y}{dx^n} + a_1\frac{d^{n-1}y}{dx^{n-1}} + \cdots + a_{n-1}\frac{dy}{dx} + a_n y = 0. \qquad (4.18)$$

If the auxiliary equation $a_0 m^n + a_1 m^{n-1} + \cdots + a_{n-1}m + a_n = 0$ has k distinct roots $m_1, m_2, \ldots, m_k$, then the general solution is

$$y = c_1 e^{m_1 x} + c_2 e^{m_2 x} + \cdots + c_k e^{m_k x}, \qquad (4.19)$$

where the c_i's are arbitrary constants.

To understand Problem 4.3 more adequately, let us recall that the equation of the normal curve with mean μ and variance σ^2 is $g(x) = e^{-(x-\mu)^2/2\sigma^2}/\sigma(2\pi)^{1/2}$; Figure 4.4 illustrates the density function of the problem for a few values of t. Dirac's function, or distribution, is defined by $\delta(x) = 0$ if $x \neq 0$, $\delta(x) = \infty$ if $x = 0$, and $\int_a^b \delta(x)\,dx = 1$ for $a < 0 < b$.

Problem 4.3. Show that

$$f(x, t/x_0, t_0) = \left[2\pi(t - t_0)\sigma^2\right]^{-1/2} e^{-(x-x_0)^2/2\sigma^2(t-t_0)}$$

satisfies the stationary diffusion equation $\partial f/\partial t = \sigma^2\,\partial^2 f/2\partial x^2$, and the initial condition $\lim_{t \to t_0} f(x, t/x_0, t_0) = \delta(x - x_0)$, where δ is Dirac's distribution. Conclude that if $f = C$ and $(x_0, t_0) = 0$, the function

$$C = k \frac{e^{-x^2/4Dt}}{t^{1/2}} \tag{4.20}$$

(k constant) satisfies Fick's one-dimensional diffusion equation.

In Problem 4.3 we have $\lim_{t \to \infty} C(t) = 0$ for all x. This is in agreement with the intuition according to which the concentration at any point x finally reaches the steady state, where it does not change any more with time. The case where $(x_0, y_0) = (0, 0)$ is known as *WIENER'S PROCESS*; Fick's diffusion equation is the simplest of all. Its solution satisfies the boundary conditions $C = \infty$ at $x = 0$ and $t = 0$, and $C = 0$ at $x \neq 0$ and $t = 0$. This solution describes the outward movement of the molecules centered at the origin where the substance is concentrated at $t = 0$. We have $\lim_{t \to 0} C(t) = 0$ for $x \neq 0$.

Problem 4.4. In the case of a cylinder with infinite length and unit area section, show that the constant k in Problem 4.3 becomes equal to $M/2(\pi D)^{1/2}$, where M is the mass of the diffusing substance.

Hint. The total quantity of diffusing substance is given by $M = \int_{-\infty}^{+\infty} C \, dx$.

Let us consider now a very large tank containing a solution whose concentration is maintained at a constant value C_0. Let us suppose that a tube is introduced through its surface. At time $t = 0$, the concentration in the tube is zero. After how long will the concentration at a point x of the tube be equal to a particular value kC_0 ($k < 1$)? In other words, if a certain concentration is reached at the distance x along the tube in the time t, what is the relation between x and t?

We use the equation $C = ke^{-x^2/4Dt}t^{-1/2}$ to estimate the distance traveled by a particle in a given time. The number of particles between x and $x + dx$ is given by $dF(x, t) = C(x, t) \, dx / \int_{-\infty}^{+\infty} C(x, t) \, dx$, where F is the distribution function. The average of the squared distances is given by $\overline{x^2} = \int_{-\infty}^{+\infty} x^2 \, dF(x, t) = \int_{-\infty}^{+\infty} x^2 C(x, t) \, dx / \int_{-\infty}^{+\infty} C(x, t) \, dx$. Replacing C by its value we obtain

$$\overline{x^2} = 2Dt. \tag{4.21}$$

This relation is valid under all the conditions where the diffusion equation applies. In the case of Crick's row of cells (Section 4.3.1), we have $L^2 = cDt$, and Crick asserts that in a biochemically realistic model, $c = 2$ is a good general value.

Problem 4.5. Verify that

$$C = k \frac{e^{-(x^2+y^2)/4Dt}}{t} \tag{4.22}$$

is a solution of Fick's two-dimensional equation in x and y. Show that $k = M/4\pi D$ for the case of an infinite surface. Verify that the corresponding three-dimensional expression is

$$C = M \frac{e^{-r^2/4Dt}}{8(\pi Dt)^{3/2}}, \tag{4.23}$$

where r is the distance to a point source in an infinite volume.

Hint. In the two-dimensional case $M = \int_{-\infty}^{+\infty}\int_{-\infty}^{+\infty} C \, dx \, dy$.

4.3.3 Solving Fick's Equation

Let us find a solution for the equation

$$\frac{\partial C}{\partial t} = 3\frac{\partial^2 C}{\partial x^2}, \qquad t > 0, \qquad 0 < x < 2,$$

under the boundary conditions

$$C(0, t) = C(2, t) = 0, \qquad t > 0,$$

$$C(x, 0) = x, \qquad 0 < x < 2.$$

We suppose the existence of a solution of the form $C(x, t) = X(x)T(t)$, where $X(x)$ and $T(t)$ are functions of x and t, respectively. We obtain $d^2X/X \, dx^2 = dT/3T \, dt$. Because the left side depends on x only, and the right side on t only, it is obvious that they are both equal to a constant, that is, $-\lambda^2$ (there is no nontrivial solution satisfying the boundary conditions if the constant is nonnegative). We thus obtain two ordinary differential equations, which can be solved by a theorem stated in Section 4.3.2 (Expressions 4.18 and 4.19): $T(t) = c_1 e^{-3\lambda^2 t}$, $X(x) = c_2 \cos \lambda x + c_3 \sin \lambda x$, and finally $C(x, t) = e^{-3\lambda^2 t}(A \cos \lambda x + B \sin \lambda x)$ is a solution of the given equation. However, according to the principle of superposition, the most general solution is an infinite sum of expressions of this type, that is,

$$C = \sum_{m=1}^{\infty} e^{-3\lambda_m^2 t}(A_m \sin \lambda_m x + B_m \cos \lambda_m x).$$

The condition $C(0, t) = 0$ gives $A = 0$. The condition $C(2, t) = 0$ gives (avoiding $B = 0$) $\lambda = m\pi/2$, $m = 0, \pm 1, \pm 2, \ldots$, and $C(x, t) = B_m e^{-3m^2\pi^2 t/4} \sin(m\pi x/2)$. Putting $t = 0$ in the last equality brings $x = B_m \sin(m\pi x/2)$, given the last boundary condition. Sums of solutions being solutions, it follows that $C(x, t) = \sum_{m=1}^{\infty} B_m e^{-3m^2\pi^2 t/4} \sin(m\pi x/2)$ is a possible solution, and $C(x, 0) = x = \sum_{m=1}^{\infty} B_m \sin(m\pi x/2)$, $0 < x < 2$.

The problem of determining the B_m's consists in finding Fourier's expansion of $f(x) = x$, for $0 < x < 2$. We then have $B_m = -4\cos(m\pi)/m\pi$, and the solution searched for is $C(x, t) = \sum_{m=1}^{\infty} (-4\cos(m\pi)/m\pi) e^{-3m^2\pi^2 t/4} \sin(m\pi x/2)$. More generally, for $0 < x < l$, $C(x, 0) = f(x)$, and for α^2 instead of 3, the solution is $C(x, t) = (2/l)\sum_{m=1}^{\infty} [\int_0^l f(x) \sin(m\pi x/l)\, dx] \sin(m\pi x/l) e^{-\alpha^2 m^2 \pi^2 t/l^2}$. Fourier showed that it is indeed possible to write a function $f(x)$, piecewise continuous, as an infinite linear combination of the functions $\sin(n\pi x/l)$ or $\cos(n\pi x/l)$ in $0 < x < l$, that is,

$$\sum_{n=1}^{\infty} a_n \sin \frac{n\pi x}{l} \quad \text{or} \quad \frac{b_0}{2} + \sum_{n=1}^{\infty} b_n \cos \frac{n\pi x}{l},$$

where

$$a_n = \frac{2}{l} \int_0^l f(x) \sin \frac{n\pi x}{l} dx, \qquad n = 1, 2, 3, \ldots,$$

and

$$b_n = \frac{2}{l} \int_0^l f(x) \cos \frac{n\pi x}{l} dx, \qquad n = 0, 1, 2, 3, \ldots$$

Problem 4.6. Find a solution $\mu(x, y)$, satisfying Laplace's equation in the rectangle $0 < x < a, 0 < y < b$, and the conditions $\mu(x, 0) = 0$, $\mu(x, b) = 0$, $\mu(0, y) = 0$, $\mu(a, y) = f(y)$.

Hint. Putting $\mu(x, y) = X(x)Y(y)$, we have $Y''/Y = -X''/X = -\lambda$ for some constant λ. The system becomes $Y'' + \lambda Y = 0$, $Y(0) = 0$, $Y(b) = 0$, and $X'' - \lambda X = 0$, $X(0) = 0$. To avoid the trivial solution, we must take $\lambda = n^2\pi^2/b^2$, which gives $Y(y) = \sin(n\pi y/b)$. Finally, $\mu_n(x, y) = \sinh(\pi n x/b)\sin(\pi n y/b)$ is a solution for $n = 1, 2, 3, \ldots$. The function $\mu(x, y) = \Sigma_{n=1}^{\infty}c_n\mu_n(x, y)$ is formally a solution, and we must choose the c_n's so that

$$f(y) = \sum_{n=1}^{\infty} c_n\sinh\frac{\pi n a}{b}\sin\frac{\pi n y}{b}, \qquad 0 < y < b.$$

In other words, we must find Fourier's expansion of $f(y)$ in terms of sines. We have $c_n = [2/b\sinh(\pi n a/b)]\int_0^b f(y)\sin(\pi n y/b)\, dy$, $n = 1, 2, 3, \ldots$.

Problem 4.7. Find solutions $C(x, y, t)$ for

$$\frac{\partial C}{\partial t} = \alpha^2\frac{\partial^2 C}{\partial x^2} + \alpha^2\frac{\partial^2 C}{\partial y^2},$$

satisfying the boundary conditions

$$C(0, y, t) = C(a, y, t) = C(x, 0, t) = C(x, b, t) = 0$$

$$\left(C(x, y, t) = \sin\frac{n\pi x}{a}\sin\frac{n\pi y}{b}e^{\alpha^2 n^2\pi^2(b^2 - a^2)t/a^2 b^2}\right).$$

4.4 PHYLLOTAXIS AND DIFFUSION

4.4.1 Diffusion Equations in Phyllotaxis

The following equation is appropriate for the domain of phyllotaxis:

$$\frac{\partial C}{\partial t} = D\nabla^2 C + P\frac{\partial C}{\partial y} - KC + Q. \qquad (4.24)$$

It combines Fick's law of diffusion, the production and decay of the inhibitor, and the transport in the direction of the Y-axis. Ordinary diffusion alone cannot account for the high rates of transport. Plants would not grow if they were not able to transfer material rapidly to places where it is needed. The transport must thus be explicitly represented in a realistic model of diffusion in phyllotaxis.

The concentration C is a function of the position vector $\mathbf{v}$ and of the time t. For Mitchison (1977) $\mathbf{v}$ is a vector (x, y, z) in three-dimensional space. For Veen (1973), Hellendoorn and Lindenmayer (1974), and Young (1978), $\mathbf{v} = (x, y)$, that is, phyllotaxis is considered essentially as a two-dimensional phenomenon; the diffusion of the inhibitor is confined to a thin slice of tissue on the surface of the shoot apex. In his model Thornley (1975a) considers that $\mathbf{v} = x$, a real number. These authors, the last one excepted, use the cylindrical representation of the primordia.

The inhibitor diffusion coefficient D is considered as a constant, K is a constant, Q is the inhibitor production rate (a function of $\mathbf{v}$ and of C), and P is the transport rate. It is generally supposed that $Q > 0$ in the primordia and in the apical meristem, and that $Q = 0$ elsewhere. The inhibitor diffuses away from its sources and is destroyed by a kinetic process at a rate proportional to its concentration (KC). As soon as C falls below a given threshold value at a point, a primordium is initiated, and Q becomes greater than zero at that point. It follows that the new primordium, as a source of inhibitor, contributes to the determination of the next primordium after a sufficient growth of the shoot apex.

Because of the complex structure of $Q(\mathbf{v}, C)$, Expression 4.24 must be solved by numerical methods. To that purpose, the apical surface is divided into zones and C is calculated, at each of these zones, for each time step. We can imagine zones of the size of a cell; the more zones there are, the more precise are the calculations, but the time required by the computer becomes enormous. Veen (1973, 1977) succeeded in doing such calculations; they are considered in Section 4.4.2. The necessity of having calculations that can be performed by the computer imposes severe constraints on the types of phyllotactic patterns that can be obtained.

In order to escape the problems of numerical analysis and the limitations due to the finite numbers of zones, and to obtain simple analytical solutions, some authors assume a steady-state inhibitor concentration, and consider, like Mitchison (1977), the linear and homogeneous equation (where C is a function of $\mathbf{v}$ only):

$$D\nabla^2 C + P\frac{\partial C}{\partial y} - KC = 0. \qquad (4.25)$$

Moreover, Thornley (1975a) and Young (1978) consider the more simple equation

$$D\nabla^2 C - KC = 0. \tag{4.26}$$

The behavior of the sources of inhibitor is now introduced as boundary conditions on the solution $C(\mathbf{v})$. We look for a solution that is maximal at the primordia and that decreases toward zero away from them.

Problem 4.8. In his diffusion theory of phyllotaxis Thornley (1975a) uses the equation

$$D\nabla^2 C(x) - KC(x) = 0,$$

where the concentration is given by unit length. Find a solution for this equation.

Answer.

$$C = Ae^{\alpha x} + Be^{-\alpha x}, \tag{4.27}$$

where $\alpha = (K/D)^{1/2}$, that is,

$$C = D\cosh \alpha x + E\sinh \alpha x, \tag{4.28}$$

and if K is negative,

$$C = F\cos \beta x + G\sin \beta x, \tag{4.29}$$

where $\beta = (-K/D)^{1/2}$.

Problem 4.9. Find a solution for the diffusion equation used by Young, in terms of polar coordinates; verify that in the rectangular coordinates

$$C(x, y) = \sum_{m=1}^{\infty} e^{\alpha_m x + \beta_m y}. \tag{4.30}$$

Hint. Use Frobenius' method (see Braun, 1978).

Problem 4.10. Determine a solution for $\partial C/\partial t = D\nabla^2 C - KC$, where C is a function of x only.

Hint. Use the transformation $C = C'e^{-Kt}$.

Problem 4.11. Find a solution for $D\nabla^2 C + P\partial C/\partial y - KC = 0$, in the case of a point source in a three-dimensional infinite volume ($C(x, y, z) = A \exp\{-[-\alpha(x^2 + y^2 + z^2)^{1/2} - \beta y]/(x^2 + y^2 + z^2)^{1/2}\}$ where $\beta = P/2D$ and $\alpha^2 = P^2/4D^2 + K/D$).

4.4.2 Derivation of the Equation for the Cellular Model

The combined equation $\partial C/\partial t = D\nabla^2 C - KC$ represents the diffusion and decay of an inhibitor in a continuous medium. Veen's equation (1973) is a two-dimensional discrete version of it. In his cellular model in phyllotaxis he considers two processes of transport: the very rapid transport inside the cell, and the diffusion through the cellular membrane. The time required for the internal diffusion is considered unimportant compared to the intercellular diffusion. Let us put $\partial C/\partial x = g_x$, and $\partial C/\partial y = g_y$, and let us suppose that the concentration inside a cell is constant. We have

$$\frac{\partial C}{\partial t} = D\left(\frac{g_{x_1} - g_{x_2}}{\Delta l} + \frac{g_{y_1} - g_{y_2}}{\Delta l}\right) - KC,$$

where g_{x_1} and g_{x_2} are the gradients on the left and right sides of the cell, respectively, g_{y_1} and g_{y_2} on the lower and upper sides of the same cell, respectively, and Δl is the diameter of the cell. Figure 4.5 shows a cell with concentration C_0, surrounded by eight cells with concentrations C_i, $i = 1, 2, \ldots, 8$, acting on C_0 in such a way that the corner cells contribute the

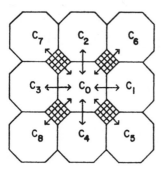

Figure 4.5 Diffusion on the cellular level in a two-dimensional tissue (from Veen and Lindenmayer, 1977).

part q in the establishment of the gradients g_{x_i} and g_{y_i}:

$$g_{x_1} = \frac{(1 - 2q)(C_3 - C_0) + q(C_8 - C_0) + q(C_7 - C_0)}{\Delta l},$$

$$g_{x_2} = \frac{(1 - 2q)(C_1 - C_0) + q(C_6 - C_0) + q(C_5 - C_0)}{\Delta l},$$

$$g_{y_1} = \frac{(1 - 2q)(C_4 - C_0) + q(C_5 - C_0) + q(C_8 - C_0)}{\Delta l},$$

$$g_{y_2} = \frac{(1 - 2q)(C_2 - C_0) + q(C_7 - C_0) + q(C_6 - C_0)}{\Delta l}.$$

It follows that

$$\frac{\partial C_0}{\partial t} = \frac{D}{(\Delta l)^2}\left[(1 - 2q) \sum_{i=1}^{4} C_i + 2q \sum_{j=5}^{8} C_j - 4C_0\right] - KC_0.$$

For reasons of efficiency and programming, Veen puts $q = \frac{1}{4}$, and obtains

$$\frac{\partial C_0}{\partial t} = \frac{D}{2(\Delta l)^2}\left(\sum_{i=0}^{8} C_i - 9C_0\right) - KC_0.$$

If the time is supposed to be discrete, $D^* = D\Delta t/2(\Delta l)^2$, and $K^* = K\Delta t$, we have

$$\Delta C_0 = D^*\left(\sum_{i=0}^{8} C_i - 9C_0\right) - K^*C_0,$$

and

$$C_0' = C_0 + \Delta C_0 = D^* \sum_{i=0}^{8} C_i + (1 - 9D^* - K^*)C_0, \tag{4.31}$$

where C_0' is the new concentration after the time interval Δt in the central cell. Hellendoorn and Lindenmayer (1974) used this formula to simulate the phyllotaxis of *Bryophyllum tubiflorum*. We return to it in Section 5.4.2.

4.4.3   Turing's Stationary Waves

According to the botanist Wardlaw (1968b), Turing's theory (1952), based on the well-known physicochemical laws, can account for certain characteristic patterns in plants, and for many facts of plant morphogenesis. The fundamental idea is to combine the effects of diffusion and chemical reactions.

Let us consider with Turing two chemical substances A and B, that is two morphogens or metabolites essential for the morphogenetic process, with concentrations C_A and C_B, respectively, in a tube. These two concentrations undergo changes due to chemical reactions. We suppose a catalyst-evocator E, that is, a substratum from which A and B can be synthesized and to which they can be degraded. There are certain values C_A^* and C_B^* of the concentrations corresponding to a chemical equilibrium. Let $C_A = C_A^* + c_A$, and $C_B = C_B^* + c_B$, where c_A and c_B measure small deviations of the concentrations from equilibrium. When $c_A = c_B = 0$, we have $\partial C_A/\partial t = \partial C_B/\partial t = 0$. For small deviations, we have (ignoring powers larger than 1, which have relatively little effect) $\partial C_A/\partial t = ac_A + bc_B$ and $\partial C_B/\partial t = cc_A + dc_B$, where a, b, c, and d are constants, the so-called marginal reaction rates. Moreover, if we have diffusion, then

$$\frac{\partial C_A}{\partial t} = ac_A + bc_B + D_A\frac{\partial^2 C_A}{\partial x^2} = a\left(C_A - C_A^*\right) + b\left(C_B - C_B^*\right) + D_A\frac{\partial^2 C_A}{\partial x^2},$$

$$(4.32)$$

$$\frac{\partial C_B}{\partial t} = cc_A + dc_B + D_B\frac{\partial^2 C_B}{\partial x^2} = c\left(C_A - C_A^*\right) + d\left(C_B - C_B^*\right) + D_B\frac{\partial^2 C_B}{\partial x^2},$$

$$(4.33)$$

where D_A and D_B are the diffusion rates of A and B. *TURING'S DIFFUSION-REACTION THEORY* asserts that under appropriate conditions, these equations determine stationary waves, that is, a regular distribution of the metabolites inside the system. Turing developed analytical solutions for his equations, in the case of a cellular ring of tissue. The interesting characteristics of their behaviors are illustrated in Figure 4.6 (from Maynard-Smith, 1968).

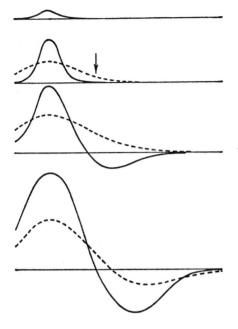

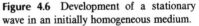

Figure 4.6 Development of a stationary wave in an initially homogeneous medium.

Let us suppose that initially $C_A = C_B = 0$ everywhere in the system and that the concentration of the catalyst-evocator E slowly increases, creating a small perturbation. Consider that a and c are positive, meaning that if C_A increases above its equilibrium level, the rates of synthesis of A and B increase. If B diffuses more rapidly than A, that is, if $D_B > D_A$, and if b is negative, that is, if the concentration of B increases, then the destruction of A follows. Figure 4.6 first shows that the equilibrium is broken by a slight increase of C_A. As C_A keeps on increasing, C_B increases, but at the point where an arrow appears, A is destroyed. This leads to the destruction of B and to the formation of a wave trough on each side of the initial peak, then to the development of further peaks, and so on until a symmetrical stationary wave settles, corresponding to a regular distribution of the metabolites.

According to Wardlaw (1968b), it is likely that this theory of standing waves will some day explain the phyllotactic systems, the branching of algae, the distribution of procambial strands in shoots, and so on. There is no proof that stationary waves constitute the foundation of forms and patterns in plants, but there is a probability that it is so. Wardlaw suggests that the birth of a primordium at a particular, and usually predictable, place is due to the activity of a system of reactions localized in the subdistal region of the shoot apex (see Figure 5.2) and to the influence of the adjacent primordia. If the preceding tube is replaced by a strip of tissue and if one of the substances, A or B, is such that its concentration, increasing above a certain threshold, brings a particular differentiation of some cells, then we would have a mechanism for explaining the regular spacing of structures in an initially homogeneous field. Unfortunately Turing died in 1954, and the article he promised on phyllotaxis, as an application of his theory, never appeared. Recently, Berding et al. (1983) produced spiral pre-patterns, resembling to those of the sunflower, by means of a diffusion-reaction mechanism, of the Meinhardt–Gierer type (1974), involving two antagonistic acting morphogens, named the inhibitor and the activator. The diffusion coeffficients are increasing functions of the distance from the pole, and the inhibitor is rapidly diffusing compared to the activator. An interesting degree of isomorphy is obtained, between the results and the real patterns; there are quantitative differences which imply the necessity of a refinement of the mechanism.

5 THE ORIGINS OF THE PATTERNS: THEORIES AND MATHEMATICAL MODELS

Theories are nets: only he who casts will catch.
(Novalis)

5.1 PRESENTATION

Many theories have been proposed to explain the phyllotactic patterns. Most of them are verbal. This chapter presents the mathematical models of plant phyllotaxis in the cylindrical, centric, and hierarchical representations. Though the investigators have been engaged in describing the phenomenon for more than a century (from Bravais and Bravais, 1837), the mathematical models inspired by these descriptions have existed for less than a decade. The present treatise reflects this historical proportion with twice as many pages devoted to the descriptive aspect of the problem. In fact, a description is a strategy for attacking a problem and for controlling its parameters and it is relatively recently (since Richards, 1948) that strategies developed coherently; whence the late formulation of the models.

These models are of four types, constituting the four main sections of the chapter: contact pressures, first available space or space filling, diffusion of an inhibitor, and systemic. The first model (Section 5.2) requires a good knowledge of Chapter 2, but the second (Section 5.3) has no prerequisite. The models of diffusion (Section 5.4) obviously require the study of Chapter 4. These are one-, two-, or three-dimensional models, in the centric or cylindrical representations, implying the use of the computer; a diffusion theory for a growing surface cannot indeed be directly formalized. In the last model (Section 5.5) it is considered that a growing shoot apex is a system where the subsystems, that is, the primordia, arise, interact in an aggregative fashion, and are set into hierarchies according to a principle of maximization of energy, that is, of minimization of entropy. In order to understand the biological foundations of this model, and the way in which the hierarchical representation of the primordia imposes upon us, the reader is referred to Research Activity 6 of Section 5.6. It is an *INTERPRETATIVE MODEL*, whereas the former are mechanistic models; it interprets the phenomenon of phyllotaxis in terms of its ultimate effect which is assumed to be the minimization of entropy. A *MECHANISTIC MODEL* is a statement about why a particular process occurs, in terms of the mechanism that is assumed to underlie that process.

These models are the object of a discussion in the Epilogue. The method of presentation of this chapter combines the methods used in the preceding chapters.

5.2 MODEL OF THE CONTACT PRESSURE THEORY

5.2.1 Contacts and Contact Pressures

The leaves against which a younger leaf is compressed are the contacts of this leaf. The spirals visually determined by consecutive contacts are called the contact parastichies; there are two families winding in opposite directions. Not all plants have contacts, in the sense of leaves compressed one on the other. In the fern, where elegant examples of normal phyllotaxis are observed, such as *Dryopteris dilatata* presenting the 8/5 pattern, the primordia of the leaves near the apex are small regions largely separated from each other (see Richards, 1948, p. 241). Nevertheless, there is something analogous to the contacts of a leaf, that is, the neighbors nearest to this leaf, measured by the distance between their centers. The analogy between the contacts and the nearest neighbors, however, can be called into question. In plants having contacts, the leaves can have the form of a rhombus or of an elongated crescent. Consequently, a contact may not be a nearest neighbor, as illustrated in Figure 5.1. Van Iterson (1907) and Richards (1951) showed how the contacts change with the form of the leaves. The concept of point of close return is more general than the notion of contact.

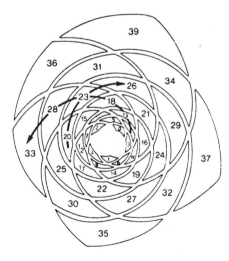

Figure 5.1 A contact of a leaf may not be one of its nearest neighbors; for example, 22 and 24 are the contacts of 27, but the center of 19 is closer to the center of 27 than the center of 24 is (from Mitchison, 1977; copyright 1977 by the American Association for the Advancement of Science.)

In 1875 Schwendener tried to give a mechanistic causal explanation of the observed divergences. He attempted to prove that the mutual pressures between the primordia were responsible for these divergences. Van Iterson also made an unsuccessful attempt in that direction. Church (1904, p. 236) criticized this theory in that initially the primordia were supposed to move under contact pressure, and the existence of such movements was not proved (and is still not). The only movement postulated in Adler's contact pressure theory (1974, 1977a) is growth, which changes the positions of the centers of the primordia so as to maximize the minimal distance between them. Even if the notion of contact appears to be a little particular, the surgical studies of Wardlaw and Cutter (1956) on the shoot apices of ferns demonstrated the mutual influences of older primordia in the control of spacing and positioning of younger members, by way of the incipient vascular tissue which permits the rapid transfer of substances; this can be called chemical contact pressures.

In Adler's model the leaves grow in diameter until they experience contact pressures; then $\delta = \Delta/C$ is maximized, Δ being the diameter of the leaf and C the girth of the base of the cylinder in the regular cylindrical representation. The analysis of the consequences of contact pressures is an analysis of the consequences of maximizing δ, the relative leaf diameter. Even if the form of the leaves changes, the pressure of growth tends to maximize the minimal distance between the centers of the leaves. Adler determined conditions for which contact pressures produce normal and anomalous phyllotaxis, by a repeated application of his maximin criterion stated below.

5.2.2   Parameters of the Model

For the purpose of determining the principle underlying the influence of contact pressures on the divergence d, Adler supposed that the divergence between the leaves n and $n + 1$ is the same for all n, and that $1/3 < d < 1/2$. This corresponds to $t = 2$ in the interval $(1/(t + 1), 1/t)$; for $t = 3, 4, \ldots$ the results can be easily extended. He supposed that during the initial period of growth of the shoot apex, the girth C of the stem increases more rapidly than the length l of the internodes, so that the rise $r = l/C$ decreases. We have seen in Chapter 3, Expression 3.59 giving r as a function of the conspicuous parastichies, that r decreases when the secondary numbers increase. In fact, any model in the cylindrical lattice that attempts to account for the observed fact of rising phyllotaxis in plants must suppose that r decreases for a while. Adler's general assumption (1977a) on r, that is,

$$\lim_{T \to \infty} r(T) = 0,$$

where T is the time in plastochrones, brought to his formulation of 1974 a simplification that confers on his model all its elegance [in Research Activity 5 of Section 3.4, this condition exists for both values of $r(T)$].

Without loss of generality, it is supposed that only one fundamental counterclockwise spiral exists. The fact that there are J fundamental spirals for whorled or multijugate plants (see Research Activity 5 of Section 2.5) simply amounts to saying that in the (d, r) plane, the length of the base of the normalized cylindrical lattice is $1/J$ instead of 1. For example, in that case, a result from Chapter 2 can be stated as follows: the opposed parastichy pair $J(S_{t, k+1}, S_{t, k})$ is visible if and only if d is equal to $F(k)/JS_{t, k}$, or to $F(k + 1)/JS_{t, k+1}$, or is between these two values.

Let us suppose now that d and r are fixed, and that leaf a is the nearest to leaf 0. We have max δ = max $d(0, a)$ = $\max(a^2 r^2 + D_a^2)^{1/2}$. It is known (see Exercise 2.1) that if r decreases, b can replace a only if $D_b < D_a$. Since $\delta = 1$ is an upper bound for δ, there is a value of d that maximizes δ. The mathematical assumption corresponding to the biological assumption of contact pressure follows.

Adler's Maximin Principle. If contact pressure begins at $T = T_c$, then for $T \geq T_c$, and for the corresponding rise $r(T)$, d has this uniquely determined value for which the distance from leaf 0 to its nearest neighbor a, in the normalized cylindrical representation, is maximized.

In the absence of contact pressure, $d(T)$ and $r(T)$ can be independent, but at $T = T_c$, for a given r, $d = d[r(T)]$. This means that because the minimal distance between the leaves is maximized, two integers p and q follow from the values of d and r, at $T = T_c$, such that (1) p and q are two neighbors of the origin, (2) p and q are on either side of the vertical axis, and (3) $d(0, p) = d(0, q)$. It follows (Chapter 2) that (p, q) is a conspicuous opposed parastichy pair, and that p and q are the denominators of consecutive principal convergents of the continued fraction of the divergence. Proposition 5 of Section 2.4.2 is then deduced, a simple criterion allowing us to determine the value of d which maximizes δ.

Maximin Criterion.   If n_{i+1} is the point of close return nearest to 0, and if $r < 1/n_{i+1}(n_{i+1}^2 - n_i^2)^{1/2}$, then for r fixed, the value of d that maximizes δ lies on the arc (n_{i+1}, n_i) of the (d, r) plane.

5.2.3   Consequences of the Model (Adler, 1974, 1977a)

The maximin principle arranges that three consecutive points of close return are linked by the recurrence relation defining the main series (see Research Activity 1 in Section 5.6, and Exercise 2.34).

Let us suppose that at the beginning of contact pressure the system is on arc (q_n, q_{n-1}). The *CONTACT PRESSURE PATH* is a sequence of arcs of circles linked to each other in the following way: first the arc (q_n, q_{n-1}) down to its intersection with the arc (q_{n+1}, q_n), where $q_{n+1} = q_n + q_{n-1}$, then the arc (q_{n+1}, q_n) from that point to its intersection with the arc (q_{n+2}, q_{n+1}), where $q_{n+2} = q_{n+1} + q_n$, and so on. The contact pressure path beginning on the arc $(t, 1)$ is the *NORMAL PHYLLOTAXIS PATH* (see Exercise 2.33). The contact pressure path beginning on the arc $(2t + 1, t)$ is one of the *ANOMALOUS PHYLLOTAXIS PATHS*.

Suppose that Q is the point of intersection of the arcs (q_{n-1}, q_n), and (q_n, q_{n+1}) in the (d, r) plane, where $q_{n+1} = q_n + q_{n-1}$, and let T_Q be the time at which the point (d, r) on the arc (q_{n-1}, q_n) reaches Q. If $q_{n+1} \le T_Q$, then, as r continues to decrease, the point (d, r) passes from one arc to the other. If $T_Q < q_{n+1}$, leaf q_{n+1} emerges after the point (d, r) passes Q on the arc (q_{n-1}, q_n). Under the condition that the minimal distance between the leaves continues to be maximized, the point (d, r) jumps horizontally from the arc (q_{n-1}, q_n) to the arc (q_n, q_{n+1}). This discontinuity in the path of (d, r) in the model springs from the assumption that the leaf q_{n+1} is fully grown at birth. In a plant, this discontinuity does not exist: the point (d, r) follows a transitional path from one arc to the other. Following are a few straightforward propositions, closely related to Proposition 6 of Section 2.4.2.

Proposition 1

For a leaf distribution where $1/(t + 1) < d < 1/t$, if at $T = T_c$ the point (d, r) is on the arc (q_n, q_{n-1}), then

$$\lim_{T \to \infty} d[r(T)] = [0; t, a_2, \ldots, a_n, 1, 1, 1, \ldots].$$

Proof. (q_n, q_{n-1}) being a visible pair, it can be reached by a sequence of extensions starting with $(0, 1)$ and passing by the visible pair $(t, t + 1)$. We have $a_1 = t$ left extensions,..., then finally a_n left extensions if n is odd, or a_n right extensions if n is even. We thus obtain the pair (q_{n-1}, q_n) or the pair (q_n, q_{n-1}). By the maximin criterion we have $a_{n+1} = a_{n+2} = \ldots = 1$.
∎

Proposition 2

A necessary and sufficient condition for contact pressure to produce normal phyllotaxis with $t = 2$ is that at $T = T_c$ the point (d, r) is confined to the arc $(F(k + 1), F(k))$ for some k.

Proposition 3

A sufficient condition for the existence of normal phyllotaxis with $t = 2$ is that either $T_c < 5$ or $r(T_c) \geq \sqrt{3}/38$; that is, this type of phyllotaxis is inevitable if contact pressure begins before leaf 5 appears or before r becomes smaller than $\sqrt{3}/38$.

Network of Vortices. Adler (1975) used the following metaphor to describe his model. The (d, r) plane contains many vortices, the contact pressure paths. In the portion of the plane where $r > \sqrt{3}/38$, there is only one vortex, and as r decreases the number of vortices increases. When contact pressure begins, wherever the point (d, r) may be in the plane, it immediately drifts toward the nearest vortex; that is, there is a change in the value of d which increases the minimal distance between the leaves of the regular lattice. Then as r decreases, (d, r) travels through the vortex, perhaps leaving it temporarily, but always coming back to it.

We have supposed that the consecutive divergences d_n are equal to d. However, on the tip of a growing stem, the d_n's may differ, although with the growth of n they tend to become alike. This disagreement between the model and the plant can be eliminated if d is identified, at time $T = i$, as the average $\sum_{j=1}^{i} d_j / i$. In fact, the hypothesis regarding the equality of the consecutive divergences can be replaced by the weaker assumption according to which the consecutive primordia lie on a single fundamental spiral. Then the value of d determined by the equation of the arc (p, q) is simultaneously equal to $\sum_{i=1}^{p} d_i / p$ and to $\sum_{i=1}^{q} d_i / q$. Moreover, since any leaf can play the role of leaf 0, d is simultaneously equal to $\Sigma_{i=1}^{p} d_{n+i-1}/p$ and to $\Sigma_{i=1}^{q} d_{n+i-1}/q$ for any n. Under this condition the following proposition shows that contact pressure forces an equalization of the d_n's between consecutive primordia.

Proposition 4

Let p and q be the two neighbors of the origin such that the point (d, r) is on the arc (p, q), where p and q are relatively prime numbers on both sides of the vertical axis. Let us suppose that d is the average of the existing d_i. Then the growth of T compels the equalization of the divergences.

Proof.   The condition that $(d, r(T))$ is on the arc (p, q) determines d as a function of r, where $d_1 + d_2 + \cdots + d_p = pd$ and $d_1 + d_2 + \cdots + d_q = qd$ (see the above paragraph). In general we have, for any n, $d_n + d_{n+1} + \cdots + d_{n+p-1} = pd$, and $d_n + d_{n+1} + \cdots + d_{n+q-1} = qd$. It follows that $d_{n+1} + \cdots + d_{n+p} = pd$, so that $d_n = d_{n+p}$. We also have $d_n = d_{n+q}$. Consequently if $m \equiv n \pmod{p}$ and if $m \equiv n \pmod{q}$, we have $d_m = d_n$. For arbitrary positive integers m and n, there is an integer k such that $m \equiv k \pmod{p}$ and $n \equiv k \pmod{q}$ (this is a particular case of the Chinese remainder theorem; see Stewart, 1966). It follows that $d_m = d_n = d_k$. With the growth of T, more and more d_i's will be equalized to the value of d determined by the condition that $(d, r(T))$ is on the arc (p, q). This continues until r is small enough for $p + q$ to replace $p < q$ as a neighbor of the origin. Then the process of equalization of the d_i's starts all over again from the value of d determined by the condition that $(d, r(T))$ moves on the arc $(q, p + q)$ (Adler, 1977a). ■

The preceding results prove Schwendener's conjecture according to which under certain conditions contact pressure replaces the phyllotaxis m/n by the phyllotaxis $(m + n)/m$; if $m = 2$ and $n = 1$, the divergence tends toward ϕ^{-2}. Schwendener supposed, without justification, that $d(0, m) = d(0, n)$ when m and n are neighbors of the origin on both sides of the vertical axis. We have just seen that this assumption is not only meant for simplifying purposes; it is a consequence of the maximin principle. Schwendener (1878) calculated the d_n's of Exercise 2.33 by an empirical method, for $t = 2$; his results are remarkably close from the former.

Schwendener's data do not reveal that contact pressures make d a function of r; it is this relation between the two parameters that opens the way to a verification of the model.

Problem 5.1. Consider the arcs $(2,1)$, $(3,2)$, $(3,1)$, $(5,1)$, $(5,2)$, and $(5,3)$. These arcs extended to their semicircles divide the plane into regions. Let us call region I the region outside the arcs $(5,2)$ and $(3,2)$, but inside the others, and region II the one that is external to the arc $(3,2)$ only. Let us define R as the rectangle whose base is the interval $(\frac{2}{5},\frac{1}{2})$ on the d axis and whose height is $1/5\sqrt{21}$ (see Exercise 2.30b). Show that contact pressures produce anomalous $(2,5)$, then $(7,5)$ phyllotaxis under the following set of conditions:

1. δ is maximized for $5 \le T_c < 6$.
2. When δ is maximized for the first time, (d, r) is in the region $R \cap (\text{I} \cup \text{II})$.
3. Leaf 7 emerges and is mature when $\sqrt{3}/78 \le r < \sqrt{7}/84$, and the next close return candidate after leaf 7 emerges when $r < \sqrt{3}/78$.

Hint. The regions I and II are the only ones where the leaves 2 and 5 are the nearest to 0. Examine the arc $(5,2)$ in order to determine the condition $r < 1/5\sqrt{21}$. Leaf 7 is the next close return candidate after leaf 5 for $d > 5/12$, from which it follows that $r < \sqrt{7}/84$. The arcs $(5,2)$ and $(7,5)$ intersect at $d = 11/26$ and $r = \sqrt{3}/78$.

5.3 MODEL OF THE FIRST AVAILABLE SPACE THEORY

5.3.1 Fundamental Postulates

The first available space, or space-filling, theory combines Hofmeister's rule with a hypothesis from Snow and Snow.

Hofmeister's Rule. Each new leaf arises in the largest gap or depression between the existing leaves surrounding the shoot apex (Hofmeister, 1868).

Snow's Hypothesis. Each new leaf arises in the first space on the growing apical cone that has reached the required minimal size and minimal distance below the very tip of the cone (Snow and Snow, 1962).

Hofmeister's rule thus indicates where the new leaf emerges, and Snow's hypothesis indicates the moment of emergence, namely, when a certain distance and a certain size are attained.

In famous experiments on *Lupinus albus*, Snow and Snow (1935) made radial cuts in the region where one of the primordia was supposed to arise, thus reducing the available space, and no primordium developed in that region. Presumably this space did not have the minimal area required for the birth of a primordium. Among the works supporting Snow's hypothesis are those of Gunckel and Wetmore (1946) on the vascular system of *Ginkgo biloba* and those of Wardlaw (1949a) on the fern *Dryopteris aristata* (see Research Activity 9 in Section 5.6).

This section is devoted to Adler's mathematical model (1975) of Snow's theory, where the terms "size" and "space" of Snow's hypothesis are given a sufficiently precise meaning so as to provide an unambiguous answer to the question regarding up to which point Snow's theory alone can account for the phenomenon of phyllotaxis.

Adler's Forbidden Zone Postulate (1974, 1977a). The region surrounding the very tip of the growing apex, where no primordium can be found, is not only a zone where the primordia cannot arise (empirical fact) but also a zone where the existing primordia cannot enter.

This postulate played a role in the determination of the maximin principle of Section 5.2.2. A leaf distribution can be represented in two ways: on the real surface (cylinder, cone, disk, paraboloid, surface of revolution) or on the normalized cylinder unfolded in the plane. The value of the divergence that maximizes the minimal distance between the leaves may differ from a surface to another. One of the reasons that motivated Adler to choose the normalized cylinder in the formulation of the maximin principle is that when only a few leaves exist on the surface of the stem, they must enter into the forbidden zone in order to be in contact; that is, contact pressure cannot exist when there are just a few leaves. Besides, on the normalized cylinder the center of the apical dome is unattainable, being sent to infinity, and the maximin principle is consistent with the postulate of the forbidden zone when contact pressure begins early.

5.3.2   Mathematical Formulation of the Postulates

Figure 5.2 represents the surface of the shoot, and the zone above the circle c at the geodesic distance s from the tip, is the forbidden zone, or Wardlaw's distal zone; it is the seat of the meristem. By an appropriate geometric transformation (see Research Activity 5 of Chapter 3), the distribution of the leaf centers on the apical cone is represented by a normalized cylinder. For all $i > 0$, let us suppose that the geodesic distances between the leaves i and $i - 1$ differ in length; let us call d_i the fraction of a turn between leaf $i - 1$ and leaf i by the shortest path, so that $d_i < \frac{1}{2}$. If the shortest path from $i - 1$ to i goes up to the right, we say that i is *on the right of $i - 1$*. Let r_i be the rise of leaf i on the cylinder, that is, the internode distance between leaf $i - 1$ and leaf i. Let us suppose that leaves $n - 2$ and $n - 1$ are the last two that have emerged.

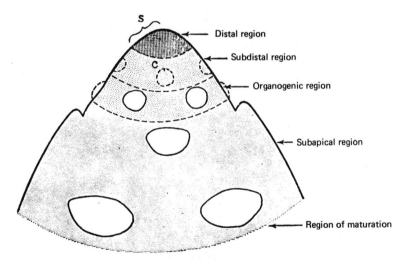

Figure 5.2   The apical meristem after Wardlaw (1957, 1968b). The primordia are initiated in the second zone, but they can be observed in the third zone only. (Copyright by Manchester University Press.)

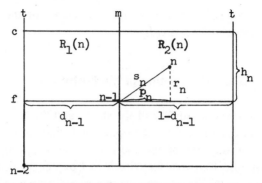

Figure 5.3 The general framework of Adler's mathematical assumptions for Snow's first available space theory. The circle c in Figure 5.2 becomes, in the unfolded normalized cylinder, a straight line orthogonal to the generator t of the cylinder, going through the leaf $n - 2$. The leaf $n - 1$ allows us to determine the regions $R_1(n)$ and $R_2(n)$.

The cylinder is unfolded in the plane, by a cut along a generator t parallel to its axis and going through leaf $n - 2$. The circle c becomes a unit segment, orthogonal to t. Let f be the line through $n - 1$ parallel to c, and m the one through the same leaf, orthogonal to c. This determines the rectangles $R_1(n)$ and $R_2(n)$ of Figure 5.3, whose common height is h_n. If $n - 1$ is on the right of $n - 2$, then $d_{n-1} < \frac{1}{2}$ and $1 - d_{n-1} > \frac{1}{2}$ are the respective bases of these rectangles. When the apical cone grows, c goes away from $n - 1$ and h_n increases. The four assumptions defining the model follow.

Assumption 1. If leaf $n - 1$ is on the right of leaf $n - 2$, then leaf n emerges inside or on the upper base of the larger of the two rectangles $R_1(n)$ and $R_2(n)$ at the moment when this rectangle attains the area $k > 0$.

This statement is consistent with Hofmeister's axiom. The numbers k and s (see Figure 5.2) represent, respectively, the minimal size and distance in Snow's hypothesis. Supposing, without loss of generality, that leaf 1 is on the right of leaf 0; then leaf 2 emerges inside or on the upper side of $R_2(2)$.

Let us now search for the weakest assumption that will imply that leaf 2 arises on the right of leaf 1. Let s_2 be the segment going up to the right from leaf 1 to leaf 2 (see Figure 5.3 with $n = 2$). Let p_2 be the horizontal component of s_2. We have $p_2 = a_2(1 - d_1)$, where $0 < a_2 < 1$, and $r_2 = b_2 h_2$, where $0 < b_2 \leq 1$. Leaf 2 is on the right of leaf 1 if and only if $p_2 < \frac{1}{2}$. If $a_2 \leq \frac{1}{2}$ then $d_2 = p_2 < \frac{1}{2}$. If $a_2 > \frac{1}{2}$, the permitted values for a_2, giving

$d_2 = p_2 < \frac{1}{2}$, are linked to the values of x such that $0 < x < d_1 < \frac{1}{2}$. In order to have $x < d_2 < \frac{1}{2}$ also, one only needs consider that $2x < a_2 < 1/2(1 - x)$. The value $x = \frac{1}{4}$ maximizes the length of the interval $[2x, 1/2(1 - x)]$ and guarantees that $a_2 > \frac{1}{2}$. The assumption sought after is thus as follows.

Assumption 2. $\frac{1}{4} < d_1 < \frac{1}{2}$ and $0 < a_2 < \frac{2}{3}$.

By induction, if leaf $n - 1$ is on the right of leaf $n - 2$, $R_2(n)$ is larger than $R_1(n)$ and leaf n emerges (Assumption 1) inside or on the upper base of $R_2(n)$. We also define s_n and p_n: s_n is the segment going up to the right, from leaf $n - 1$ to leaf n, $p_n = a_n(1 - d_{n-1})$, and $r_n = b_n h_n$, where $0 < a_n < 1$ and $0 < b_n \leq 1$. To simplify the model, Adler states the following assumption.

Assumption 3. $a_n = a$ for all $n \geq 2$.

The following assumption is weaker.

Assumption 3*. When n is large enough the consecutive values of a_n are approximately equal.

From the first three assumptions we can deduce Proposition 1 of the following section. From Proposition 2 of the same section, notice that the consecutive rectangles $R_2(n)$ tend to become congruent as n increases, so that it is reasonable to impose the following assumption.

Assumption 4. $\lim_n b_n = b$.

The distance from the center of a new leaf to c is given by $h_n - r_n = (1 - b_n)k/(1 - d_{n-1})$. The sequence $\langle h_n - r_n \rangle$, $n \geq 1$, possesses a greatest lower bound $e \geq 0$. Let C be the circle parallel to c at the distance e from c, under c. The circle C, uniquely determined, is among all the possible circles the one that is farthest from the tip of the shoot apex.

5.3.3 Consequences of the Model (Adler, 1975)

Following is a set of propositions deducible from the preceding assumptions. The proof of the first proposition easily follows from an induction on n for both cases $a \leq \frac{1}{2}$ and $\frac{1}{2} < a < \frac{2}{3}$.

Proposition 1

In the model defined by the first three assumptions of Section 5.3.2, leaf n is on the right of leaf $n - 1$ for all $n \geq 1$, and the consecutive leaves are on a genetic spiral that goes up to the right.

Proposition 2

The series $\langle d_n \rangle$, $n \geq 1$, converges toward $d = a/(a + 1)$; the series $\langle r_n \rangle$, $n \geq 1$, converges toward $r = bk(a + 1)$. It follows that the genetic spiral tends to become a regular helix, on which the leaves are equidistant.

Proof. The recurrence equation $d_n = a(1 - d_{n-1})$ gives $d_n = a - a^2 + \cdots + (-1)^n a^{n-1} + (-1)^{n-1} a^{n-1} d_1$, that is, $d_n = (a + (-1)^n a^n)/(a + 1) + (-1)^{n-1} a^{n-1} d_1$. Since $a < 1$, $\lim_n a^n = 0$, and $\lim_n d_n = a/(a + 1)$. From Assumption 1, $h_n = k/(1 - d_{n-1})$ for $n \geq 2$, so that $\lim_n h_n = k(1 + a)$. Finally $r_n = b_n h_n$ for all n, and Assumption 4 implies that $\lim_n r_n = bk(a + 1)$. ∎

Proposition 3

If Assumption 3* is substituted for Assumption 3, then the model can explain why the leaves tend to become equidistant on a helix as n increases.

Proof. From the relation $d_n = a_n(1 - d_{n-1})$, we have $d_n = a_n - a_n a_{n-1} + \cdots + (-1)^n \pi_{i=2}^n a_i + (-1)^{n-1} d_1 \pi_{i=1}^n a_i$. Since the convergence exists when $d_n = a(1 - d_{n-1})$, then for n large enough, and for every positive integer j, the values $d_n, d_{n+1}, \ldots, d_{n+j}$ are approximately equal. Given that d_n is a continuous function of $a_1, a_2, \ldots, a_n$, for all n, the resulting values for $d_n, d_{n+1}, \ldots, d_{n+j}$ are, for n large enough, approximately equal to the preceding values if $a_2, a_3, \ldots, a_n$ are sufficiently close to a. ∎

Proposition 4

The model defined by Assumptions 1–4 cannot explain the convergence of the d_n's toward the angle ϕ^{-2}, unless the gratuitous assumption $a = \phi^{-1}$ is made.

Proof. From $0 < a < \frac{2}{3}$, we get $0 < d < \frac{2}{5}$. Unless supplementary restrictions are laid on a, any value for d between $0°$ and $144°$ is possible. If $a = \phi^{-1}, d = \phi^{-2}$. ∎

Notice that a model using as strong an assumption as $a = \phi^{-1}$ would assume the regulation of the phyllotactic systems instead of accounting for it. The preceding result confirms Snow and Snow's perception (1962) according to which "...the serious difficulty has remained that it has not been clear how a theory of this kind can account for the exact regulation of phyllotaxis systems to their usual fairly high accuracy...".

Proposition 5

If $\frac{1}{2} < a < \frac{2}{3}$, the phyllotaxis 2/1, 3/2, and 5/3 are possible.

Proof. For these values of a, we have $\frac{1}{3} < d < \frac{2}{5}$. From Section 2.2.2 the points of close return are 1, 2, and 3, and the next close return candidate is 5. So for appropriate values of the rise r we have the phyllotaxis pointed out. ∎

5.4 MODELS OF THE DIFFUSION THEORY

5.4.1 One-Dimensional Model (Thornley, 1975a)

5.4.1.1 Generation of the Equations. Thornley's model (1975a, 1976) concerns the diffusion of a morphogen whose nature stays undecided. The primordia are considered as points on a circle, and the morphogen they produce diffuses tangentially to it. A primordium arises where the morphogenetic field is minimal. Figure 5.4 illustrates the system of coordinates in use. The model has two parameters, α and λ. The equation used is $d^2C/dx^2 - \alpha^2 C = 0$ (Expression 4.26), where $\alpha = (K/D)^{1/2}$; the general solution is $C = Ae^{\alpha x} + Be^{-\alpha x}$, A and B being the constants resulting from the boundary conditions. As for boundary conditions Thornley assumes that $C(x = 0) = C(x = 2\pi c)$, and that $S_1/2 = -D\,dC/dx$ at $x = 0^+$, S_1 being the strength of source P_1 (see Expression 4.1).

Problem 5.2. Show that the boundary conditions imply that C_1, the concentration related to S_1, is given by $C_1(x) = S_1 \cosh[\alpha(x - \pi c)]$ $/2D\alpha \sinh(\alpha\pi c)$. Calculate the mean value $\overline{C}_1$.

Hint. $\overline{C}_1 = \int C_1 dx / \int dx$ (ans.: $S_1/2\pi cK$).

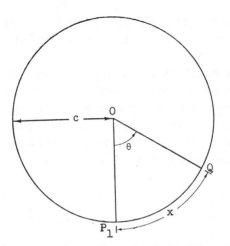

Figure 5.4 Point 0 represents the center of the shoot apex, and the new primordia are initiated at the constant geodesic distance c from 0 (the apex does not change in size). The first primordium, located at P_1, is used as a fixed mark for the determination of θ, the angular coordinate of point Q.

Problem 5.3. Determine the minimal and maximal values of C_1, and study the effect of the variation of c on these values.

Hints. $C_1(\text{min}) = S_1/2D\alpha \sinh(\alpha\pi c)$, $C_1(\text{max}) = S_1/2D\alpha \, \text{tgh}(\alpha\pi c)$; $C_1(\text{min})$ approaches zero, and $C_1(\text{max})$ approaches $S_1/2D\alpha$ as c tends to infinity. Problem 5.2 shows that if c increases, the average concentration of the morphogen decreases, and the births of the primordia arise sooner, or happen at a quicker rhythm.

The equation used being linear, if S_1 and S_2 are two sources of morphogen with respective concentrations C_1 and C_2, then $C_1 + C_2$ is the morphogenetic field created by $S_1 + S_2$, each source bringing its own independent contribution. If the source S_i is located at the position x_i, then we easily have $C_i(x) = s_i\cosh[\alpha(x - x_i - \pi c)]$ in $x_i \leq x < 2\pi c$, $C_i(x) = s_i\cosh[\alpha(x - x_i + \pi c)]$ in $0 \leq x < x_i$, and finally $dC_i(x)/dx = s_i\alpha \sinh[\alpha(x - x_i \pm \pi c)]$, where, for simplifying purposes, $s_i = S_i/2D\alpha \sinh(\alpha\pi c)$. Given the nature of the model, the second primordium arises at $x_2 = \pi c$, where $\theta = 180°$. The resulting field is, in $\pi c \leq x \leq 2\pi c$, $C(x) = s_1\cosh[\alpha(x - \pi c)] + s_2\cosh[\alpha(x - 2\pi c)]$ (the field is symmetrical about $x = \pi c$). Notice that the field depends on α and on $\lambda = S_1/S_2$. The parameter λ describes the relative strength of two primordia considered as sites of morphogen production and whose ages differ by one plastochrone. This parameter is a conglomerate of quantities representing many aspects of the problem on which there is nearly no information. The morphogenetic field coming from the first three primordia is given by $C(x) = s_1\cosh[\alpha(x - \pi c)] + s_2\cosh \alpha x + s_3\cosh[\alpha(x - x_3 - \pi c)]$, in $0 \le x < \pi c$.

Problem 5.4. Show that the point x_3, site of the minimum of the field settled by the sources S_1 and S_2, is given by

$$\text{tgh}(\alpha x_3) = \frac{\lambda \sinh(\alpha \pi c) + \sinh(2\alpha \pi c)}{\lambda \cosh(\alpha \pi c) + \cosh(2\alpha \pi c)},$$

where $\pi c \le x_3 < 2\pi c$ and $\lambda = S_1/S_2$. Show that $x_3 = 2\pi c(1 + \lambda/2)/(1 + \lambda)$ when α tends toward zero, and $x_3 = 3\pi c/2$ when α tends toward infinity and $\lambda \ne 0$. Show, lastly, that the position of the fourth primordium is given by

$$\text{tgh}(\alpha x_4) = \frac{s_1\sinh(\alpha \pi c) + s_2 + s_3\sinh[\alpha(x_3 - \pi c)]}{s_1\cosh(\alpha \pi c) + s_2 + s_3\cosh[\alpha(x_3 - \pi c)]},$$

where $0 \le x_4 < \pi c$.

The computation of the position of the first four primordia generates an inductive process by which the other primordia can be positioned. For j primordia we have $C(x) = \Sigma_{i=1}^{j}s_i\cosh[\alpha(x - x_i - f_i\pi c)]$, where $f_i = \pm 1$ depending on the interval selected, and if x' is a minimum for C, then $\text{tgh}(\alpha x') = \Sigma_{i=1}^{j}s_i\sinh[\alpha(x_i + f_i\pi c)]/\Sigma_{i=1}^{j}s_i\cosh[\alpha(x_i + f_i\pi c)]$. The field can have many minima (different values for x'); the smallest is chosen for the site of the $(j + 1)$th primordium. There are in fact j minima, one for each of the j intervals. It is also assumed that two consecutive sources have a relative strength equal to λ, and that $s_j = 1$, $s_{j-1} = \lambda, \ldots, s_1 = \lambda^{j-1}$. This means that as the time goes by the first sources weaken and have less influence on the formation of the new primordia.

Problem 5.5. Show that for $\partial C/\partial t = D\, \partial^2 C/\partial x^2 - KC$, and for a morphogenetic source $S_i e^{-ut}$ located at $x = x_i$, we have $C_i(x, t) = s_i e^{-ut}\cosh[\alpha(x - x_i - f_i\pi c)]$, where $\alpha = [(K - u)/D]^{1/2}$.

5.4.1.2   Generation of the Primordia, Simulation. Considering Problem 5.3, notice that if $\lambda = 0$, the primordium arises at $x = 0$; that is, the system produces opposite primordia. With $\lambda = 1$, we have two equal minima located at $x = \pi/2$ and at $x = 3\pi/2$, so that the third primordium arises at one of these two positions. Table 5.1 shows the relative importance of the parameters α and λ. The results of Problem 5.4 are illustrated in it. When α is very large, the sources S_1 and S_2 behave as though they were equal. There are two possible positions for x_3, because of the symmetry with regard to $x = \pi c$.

Table 5.2 was generated by the computer for various values of λ (first column) and for $\alpha = 5$. It gives the angles θ_i between the primordia i and $i + 1$, $i = 1, 2, \ldots, 11$. Thornley certifies that for $\lambda = 0.8$, the angles do not show any limiting tendency, even after the birth of the 24th primordium. It is important to note that this generation of the consecutive primordia does not always give a constant divergence angle, or a divergence that settles relatively rapidly around some value (as for the first two rows in Table 5.2). That is why Thornley proposed two variants of his deterministic model, which can be summarized physiologically as follows. A primordium arises when the morphogenetic field falls below a given threshold value, determined by the smallest among the minima of the field (we have a variable threshold). When the primordium comes into existence, it becomes a source of the morphogen, and consequently the morphogenetic level is lifted up, thus

Table 5.1   Computing the angle θ_2 between the primordia P_2 and P_3 for various values of λ and of α (from Thornley, 1975a, 1976)

	α							
λ	0	0.2	0.5	1.0	2.0	5.0	10.0	∞
0	180	180	180	180	180	180	180	—
0.01	178.2	178.2	178.2	178.1	177.9	175.9	163.6	90
0.1	163.6	163.6	163.5	163.1	161.6	151.6	130.3	90
0.2	150.0	150.0	149.8	149.3	147.3	136.2	118.4	90
0.5	120.0	120.0	119.9	119.5	118.0	110.9	102.3	90
0.7	105.9	105.9	105.8	105.6	104.7	100.9	96.3	90
0.9	94.7	94.7	94.7	94.6	94.4	93.2	91.9	90
1.0	90	90	90	90	90	90	90	90

Table 5.2 Consecutive divergence angles between the first 12 primordia, for $\alpha = 5$, and for five values of λ (from Thornley, 1976)

λ	1–2	2–3	3–4	4–5	5–6	6–7	7–8	8–9	9–10	10–11	11–12	Limiting angle
0.1	180.0	151.6	159.2	157.5	157.9	157.8	157.9	157.9	157.9	157.9	157.9	157.9
0.2	180.0	136.2	152.8	148.4	149.7	149.5	149.4	149.5	149.5	149.5	149.5	149.5
0.5	180.0	110.9	156.0	141.5	149.6	133.2	154.7	130.7	140.6	140.8	138.8	139.9
0.7	180.0	100.9	165.1	139.5	160.1	118.0	168.4	117.3	138.2	141.3	134.9	138.2
0.8	180.0	96.8	170.1	138.2	166.7	108.8	174.2	244.3	222.6	202.5	210.8	Unstable

preventing the immediate inception of other primordia at the preceding minima. A new steady-state equilibrium settles, new primordia arise, and the process starts all over again.

First Variant. One of the variants proposed by Thornley consists in introducing small perturbations in the preceding system, in order to single out a minimum other than the smallest one. Table 5.3 illustrates this situation for certain values of the parameters. In (a) the smallest minimum is chosen, as previously. In (b) a small irregularity is assumed to arise just before the inception of the eighth primordium. In nature the angle of 137.5° does not result from acrobatics or from a random stratagem; that is probably why Thornley introduced a second variant of the initial model.

Second Variant. Thornley now assumes a constant divergence angle β, corresponding to $\Delta x = c\beta$, and he considers a large number (100) of primordia separated by this angle. He also supposes that $s_{j-1}/s_j = \lambda$. Using a computer to calculate the field resulting from his equations, he finds the

Table 5.3 Positioning the primordia with $\alpha = 2$, and $\lambda = 0.7$. In (a) the smallest among the minima is chosen; in (b) the eighth primordium arises at a minimum adjacent to the smallest among them (from Thornley, 1975a)

	1–2	2–3	3–4	4–5	5–6	6–7	7–8	8–9	9–10	10–11	11–12	Limiting angle
(a)	180.0	104.7	163.2	140.5	159.7	120.5	165.3	149.9	147.6	150.7	148.5	148.8
(b)	180.0	104.7	163.2	140.5	159.7	120.5	135.6	142.9	138.9	137.9	139.6	139.2

smallest among the minima, and the position of the 101st primordium. "If the angle between P_{100} and P_{101} is equal to the constant divergence angle β originally assumed, then β must be one of the required solutions to the problem" (1976, p. 258). This variant also consists in introducing small perturbations. He builds a table similar to Table 5.1, for a few values of α and of λ and for a dozen perturbations; it can be summarized as follows: (1) the angles obtained lie in the bands 78–85, 97–101, 105–110, 132–134, 136–143, and 146–180°; (2) the angle ϕ^{-2} is approached for values of λ near 1, for the five values considered for α, and for a given perturbation among the many others; (3) when $\lambda > \frac{1}{2}$ only the angle ϕ^{-2} is possible; (4) the angles of 99.5° and of 151.1°, corresponding to the sequences $1, 3, 4, 7, 11, 18, 29, \ldots$ and $2, 5, 7, 12, 19, 31, \ldots$ belong to the regions given in (1), and can arise for large values of α and of λ; and (5) the angles of 77.95° and of 64° corresponding to other sequences observed in nature $(1, t, t + 1, 2t + 1, \ldots, t = 4, 5)$ are not in the regions mentioned in (1). The equation used by Thornley is valid in the steady-state case where the morphogenetic source has a constant strength. However, it is likely that the rate of production of the morphogen varies with time.

Spatial Competence. Thornley's model describes more or less happily spiral, distichous, and spirodistichous phyllotaxis. He introduces the notion of spatial competence to describe the situations where the primordia arise in groups: there must be a sufficient quantity of competent tissue for a primordium of a given minimal size to exist. The model is then able to explain decussate, bijugate, and some forms of whorled phyllotaxis. It fails to explain opposite pairs of leaves lying in a single vertical plane, and the whorled patterns where the leaves in adjacent whorls are vertically above one another. In order to explain all the cases, Thornley proposes to modify his model where the repulsive properties are made attractive.

Whorled Phyllotaxis. Here is how Thornley explains the whorled patterns where there are many leaves at the same level of the stem (in fact, the mechanism supposes that the leaves in a whorl are not all at the same level). After the inception of a primordium, the morphogenetic field continues to fall, thus allowing the consecutive, rapid, and discrete births of the other primordia of the same whorl (this is in opposition to a preceding assumption according to which the field increases from the instant of the birth of a primordium). As soon as the whorl is in place, spatial incompetence prevents, for a relatively long period, the inception of new primordia. Because of the recently born primordia, the field increases, climbs beyond the threshold, and the apex continues to grow until it reaches a size permitting primordial initiation. But then the morphogenetic field prevents them from arising. When the threshold is attained again, the apex has the needed competence and the mechanism starts all over again.

Thornley concludes that it would be more satisfying to deduce ϕ^{-2}, with precision, from nonnumerical methods, and that the idea of spatial competence supports Snow and Snow's hypothesis (Section 5.3.1).

5.4.2 Cellular Model (Veen, 1973; Veen and Lindenmayer, 1977)

5.4.2.1 Simulation. Section 4.4.2 presents the equation used by Veen and Lindenmayer to simulate the phyllotactic patterns. This model is given in Veen's doctoral thesis (1973). The inhibitor is produced by the centers of the leaves and by the upper disk of the cylinder; it diffuses isotropically (in all directions) and decays at a constant rate, thus regulating phyllotaxis. The cells are laid out in rows on the cylinder unfolded in the plane, and the concentration inside each cell is uniform. At each (discrete) step the inhibitor concentration is deduced from Expression 4.31. Each cell where the concentration falls below some fixed threshold S (variable in the preceding model) becomes the center of a leaf where the concentration of the inhibitor is maintained at the constant value L. The upper row of cells also maintains a constant concentration A. To simulate growth, at every G steps a new upper row of cells, each having the concentration A, is added. This parameter G controls the rate of inception of the new leaves, and the other parameters control their positions. After each step the cells where the concentration is smaller than S suddenly become cells with concentration L.

This model has been programmed in Algol 60 from an initial configuration: there are w columns of cells on the surface of the cylinder ($2\pi c$ in the preceding model) and R rows of cells, the cells (those of the upper row excepted) have the uniform concentration B, and an initial distribution of leaves is defined (with concentration L). The first question that arises is to determine whether it is possible to give values to parameters of the theory so that the initial distribution will be reproduced.

Figure 5.5 illustrates the program worked out by the computer. It gives a positive answer to the inquiry formulated in the last paragraph. The concentration in a cell is denoted by one of the symbols in Table 5.4. The

```
1    8  555555555555555555555555      1800   59  5555555555555555555555555
     7  444444444444444444444444             58  333333333333333333333333
     6  444444444444444444444444             57  222222222222222222222222
a    5  444444444444444444444444             56  111111111111111111111111
     4  444444444444444444444444             55  111111111111111111111111
     3  AA4444444444444444444444A            54  111111111111111122222111
     2  ■A4444444444444444444444A            53  110000111111111111223332
     1  555555555555555555555555             52  110001111111111224666421
                                             51  111111111222212236CDC6321
                                             50  111111122333222237D■D7321
                                             49  111111224666432236CDC6322
36   9  555555555555555555555555             48  222211236CDC6322346764322
     8  444444444444444444444444             47  333322237D■D7422233433323
     7  333333333333333333333333             46  676432236CDC6322223322234
b    6  333322222222222222222333             45  CDC63223467643222322223
     5  443222222222222222222234             44  D■D74222334333333343322247
     4  764322222222222222222346             43  CDC63222233222346764322236
     3  DB533222222222222222235B             42  67643222332222246CDC632234
     2  ■864333333333333333333346B           41  343333334332224 7D■D742223
     1  555555555555555555555555             40  3322234676432236CDC642222
                                             39  3222246CDC632234676432223
                                             38  3322247D■D742223343333334
100  10 555555555555555555555555      f      37  6432236CDC6422223322234 67
     9  333333333333333333333333             36  C63223467643222332222 4DCD
     8  222222222222222222222222             35  D74222334333333343322247D■
c    7  222211111111111111111222             34  C64222233222346764322236CD
     6  222211111111111111111222             33  643222332222246CDC63223467
     5  333211111111111111111233             32  33333343322247D■D74222334
     4  654321111111111111112345             31  22234676432236CDC642222233
     3  DB532222222222222222235B             30  22246CDC63223467643222332
     2  ■864333333333333333333346B           29  22247D■D742223343333334 33
     1  555555555555555555555555             28  32236CDC642222332222 46764
                                             27  32234676432223322224 6CDC6
                                             26  42223343333334332224 7D■D7
252  15 555555555555555555555555             25  42222332222346764322236CDC6
     14 333333333333333333333333             24  32223322224 6CDC6322346/64
     13 222222222222222222222222             23  33334322224 7D■D7422233322
     12 111111111111111111111111             22  234676432236CDC642222332 2
     11 111111111111111111111111             21  246CDC63223467643222332 2
     10 111111111111111111111111             20  247D■D74222334333333433 22
d    9  111111122222111100000111             19  236CDC6422223322234676432
     8  111111223332211100000111             18  2346764322232223222 46CDC632
     7  111112346664211110000111             17  22334333334332224 7D■D742
     6  22212236CDC63211111111112            16  222332223467643223 6CDC642
     5  33222247D■D74211111111223            15  2233222246CDC63223467643
     4  65432346CEC64221111112245            14  3343332224 7D■D74222334333
     3  DB5333457775432222222235B            13  4676432236CDC642222332223
     2  ■8644444555544333333333346B          12  6CDC632234676432223322224
     1  555555555555555555555555             11  7D■D743232344333344332224
                                             10  6CDC642222332223467643223
                                             9   4676432223322246CDC63223
                                             8   33443333444322347D■D74222
                                             7   2332223467643223 6CDC64222
                                             6   2222346CDC643234676432 22
                                             5   43322347E■E74322344433333
                                             4   66433346CEC643222333333346
                                             3   DB5433457875433223333345B
                                             2   ■86444445555443333333346B
                                             1   555555555555555555555555
```

Leaves arose in

Block	Row	Column	
0	2	1	input
105	5	10	input
253	8	19	
341	11	3	
446	14	12	
551	17	21	
656	20	5	
761	23	14	
866	26	23	
971	29	7	
1076	32	16	
1181	35	25	
1286	38	9	
1391	41	18	
1496	44	2	
1601	47	11	
1706	50	20	

```
254  15 555555555555555555555555
     14 333333333333333333333333
     13 222222222222222222222222
     12 111111111111111111111111
     11 111111111111111111111111
     10 111111111111111111111111
     9  111111122222111100600111
e    8  111111223332211106■600111
     7  111112346664211166601111
     6  22212236CDC63211111111112
     5  33222247D■D74211111111223
     4  65432346CEC64221111112245
     3  DB5333457775432222222235B
     2  ■8644444555544333333333346B
     1  555555555555555555555555
```

Figure 5.5 Generating phyllotaxis 3/2 by the diffusion of an inhibitor on the surface of a cylinder (from Veen and Lindenmayer, 1977). At the 1800th step there are 17 leaves.

values of the parameters are $K = 0.1$, $D = 0.111$, $L = 50$, $B = A = 5$, $S = 0.4$, $G = 35$, $R = 8$, and $w = 25$ (for various reasons, such as the limits of the computer, $w \leq 50$). Two leaves, the two black squares at step d in Figure 5.5, have been placed at predetermined positions; it is an input given by the horizontal and vertical distances, $u = 9$ and $v = 3$, from leaf 1 to leaf 2 introduced at the 105th step of the process. It is easily calculated that the initial divergence $2\pi u/w$, the distance between two leaves, and the rise $2\pi v/w$ are respectively given by 2.262 or 129.6°, 2.318, and 0.754. The computer rounds up to the nearest integer the concentration in each cell. The initial pattern is reproduced, and the phyllotaxis 3/2 follows, since leaves 3 and 4 are leaf 1's nearest neighbors.

5.4.2.2 Results. Table 5.5 shows the 20 patterns resulting from the previous conditions. Multiple values for the parameters w, u and v give the same pattern. For example, the values $w = 26$, $u = 10$, and $v = 2$ and the values $w = 13$, $u = 5$, and $v = 1$ yield the other parameters, phyllotaxis 3/2, and the same value for d, that is, 138.461°. To generate the arrangement 5/3, a minimum of five leaves had to be laid out before the beginning of the simulation. Moreover, the "*erase option*" had to be used: if many cells fall below the threshold S at the same time, only the cell where the concentration of the inhibitor is the smallest becomes a leaf. With this mechanics we can theoretically generate the multijugate patterns $J(m, n)$ by setting at the beginning of the operation, symmetrically in a given row of cells, Jn leaves. This model, as well as the preceding one, demonstrates the importance of the increase in diameter of the apex with regard to the pattern obtained. In the former the possible relations between the parameters of the simulation

Table 5.4 The symbols used by the computer for representing the numbers of units of the inhibitor, from 0 to 80, concentrated in the cells. For example, ? represents 69 units (from Veen and Lindenmayer, 1977)

```
0123456789ABCDEFGHIJKLMNOPQRSTUVWXYZ:ABC
|         |         |         |
0         10        20        30

DEFGHIJKLMNOPQRSTUVWXYZ:+-*/;    = <(>[])]v^
|         |         |         |         |
40        50        60        70        80
```

Table 5.5   Results of the simulation: 20 types of phyllotaxis; s is the minimal distance between the leaves, and d is the divergence angle (in degrees) (from Veen and Lindenmayer, 1977)

w	u	v	(m, n)	d	s
5	2	1	(1, 2)	144.000	2.8099
10	3	1	(1, 3)	108.000	1.9869
13	5	1	(2, 3)	138.461	1.7426
15	4	1	(1, 4)	96.000	1.7271
15	7	4	(1, 2)	168.000	3.3771
17	4	1	(1, 4)	84.706	1.5239
24	5	1	(1, 5)	75.000	1.3349
25	7	1	(3, 4)	100.800	1.2566
25	9	3	(2, 3)	129.800	2.3171
26	5	1	(1, 5)	69.231	1.2322
29	12	1	(2, 5)	148.965	1.1668
34	13	1	(3, 5)	137.647	1.0776
35	6	1	(1, 6)	61.714	1.0920
37	6	1	(1, 6)	58.378	1.0329
39	14	5	(1, 2)	129.231	2.3950
40	11	3	(1, 3)	99.000	1.7910
41	9	1	(4, 5)	79.024	0.9813
48	7	1	(1, 7)	52.500	0.9256
50	7	1	(1, 7)	50.400	0.8886
50	17	6	(1, 3)	122.400	2.2654

and the physical parameters are analyzed; some aspects of this simulation do not correspond to the observations. Notice that there is a correspondence between the parameters used in this model and those of Thornley's model, except for the parameter λ introduced by the latter.

Phyllotaxis can change under the influence of chemical agents (see Research Activity 7 in Section 3.4), microsurgery (Wardlaw, 1949; Snow and Snow, 1935; Cutter and Voeller, 1959; Loiseau, 1969), and ionizing radiation (Gunckel, 1965); it "is dependent on circumstances and, if interfered with, may readily change from one system to another" (Richards, 1948, p. 219). There are natural systems with variable phyllotaxis. For example, the mature *Bryophyllum tubiflorum* shows not less than five different patterns. Hellendoorn and Lindenmayer (1974) simulated the growth of this plant by the gradual variation of L, decreasing from 122 to 12 (see Research Activity 5 in Section 5.6).

5.4.3 Steady-State, Two-Dimensional Model (Young, 1978)

This model in which phyllotaxis is considered as a two-dimensional phenomenon simulates the diffusion of an inhibitor with steady-state concentration, in the cylindrical representation. Since $C(x, y) = X(x)Y(y)$, the method of separation of the variables (Section 4.3.3) applied to the equation used (Expressions 4.26 and 4.30), gives, for n primordial sources and the cylindrical boundary conditions,

$$C_P(x, y) = A \sum_{i=1}^{n} \cosh\left[B\left(|x - x_i| - \frac{x_L}{2}\right)\right] e^{-E|y - y_i|},$$

where $B = (K/2D)^{1/2} = E$, and A are adjustable parameters, x_L is the girth of the cylinder, and (x_i, y_i) is the site of the ith primordium. Young supposes that the apical meristem acts as a source diffusing the inhibitor, concentrically and continuously. The contribution of the meristem to the concentration of the inhibitor at the point y on the cylinder is assumed to be equal to

$$C_M(y) = A_M e^{-E_M(y_L - y)},$$

where A_M and E_M are adjustable parameters and y_L is the location of the meristem. The total concentration at the point (x, y) is thus

$$C(x, y) = C_P(x, y) + C_M(x, y).$$

Finally, Young puts $A = x_L = 1$, $A_M = 2\cosh(B/2)$, and $E_M = 1$, the last two values being chosen so as to preserve the forbidden zone (Section 5.3.1).

Problem 5.6. Show that $C_{max} = \cosh(B/2)$ and that C_{min} at $y = y_i$, due to one source only, is always equal to 1.

The model proposed by Young obviously differs from the preceding two, but requires the use of a computer as they do. The process assigns a value to

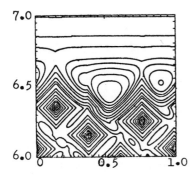

Figure 5.6   Simulation of 3/2 phyllotaxis. Leaf 0 is at $x = 0$, and $x = 1$; $y_L = 9.606$. A new primordium is arising in the middle of the figure, and a second one on the right. [From Young, 1978; with permission from the *Journal of Theoretical Biology*, **71**, 1978, p. 425. Copyright: Academic Press Inc. (London) Ltd.]

B, and a value to the threshold C_T of the concentration, positions two initial sources of the inhibitor (the two cotyledons), and fixes the initial distances between these points and the meristem at $y = y_L$. The computer then comes into action to determine the exact position of the minimal value of $C(x, y)$ in the region between the youngest primordium and the meristem. The value of this minimum is then compared to C_T, and an iterative process adjusts y_L until $C_{min} \simeq C_T$. A new source of inhibitor is then placed at the point obtained, and the mechanism is repeated. Figure 5.6 is one of its results.

The idea is obviously to obtain approximately equal consecutive divergences. For the process to become convergent, an initial distribution of 15–50 primordia is needed, depending on the value of C_T. Figure 5.7 illustrates that convergence.

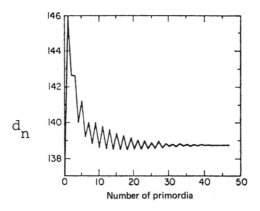

d_n

Number of primordia

Figure 5.7   The progressive equalization toward 138.74° of the consecutive divergences d_n for the phyllotaxis 3/2; $B = 14$ and $\lambda = 0.07$. [From Young, 1978; with permission from *The Journal of Theoretical Biology*, **71**, 1978, p. 426. Copyright: Academic Press Inc. (London) Ltd.]

Young asserts that for small values of $\lambda = C_T/C_{max}$, the convergence of the d_n's is very fast toward 180°, giving a spirodistichous 1/1 pattern. When λ increases the d_n's are rapidly equalized toward 137.5°, giving the patterns 2/1, 3/2, and 5/3. For patterns expressed by larger secondary numbers, the convergence is very slow. It can be quickened by placing the two initial primordia at 137.5° from one another, or by considering only the 10–15 latest primordia. For $B = 14$, and $\lambda \leq 0.2$ the initial decussate conditions produce a gradual convergence toward Fibonacci's patterns. For $\lambda \geq 0.2$, a spirodecussate system is obtained, where the d_n's oscillate between a value near 90° and another one around 180°. As λ increases, the bijugate patterns 2/2, 4/2, and 6/2 arise. The values $B = 14$, $\lambda \geq 0.7$, and three initial primordia give the pattern 3/3, and for $\lambda = 1.1$ the pattern 6/3 follows. For $\lambda = 0.73$ and $B = 14$, 5/3 phyllotaxis is displayed. Thus the pattern noticeably depends on the initial conditions and many lasting patterns can arise for any given value of λ. Young asserts (personal communication, August 1982) that he obtained, in the centric representation, normal phyllotaxis ($t = 2$) with secondary numbers higher than 3 and 5. Finally he studies the subpatterns of axillary buds, with the help of a second inhibitor, produced by these buds.

5.5 MODEL OF THE PHYLETIC THEORY

5.5.1 Fundamental Assumptions

5.5.1.1 Hierarchical Representation.
The preceding models work at the level of primordial ontogenesis, at the surface of the plant apex. But according to many botanists, such as Church (1920) and Corner (1981), phyllotaxis is first of all a phyletic problem; the morphogenesis of brown algae would hold the key to the problem. These plants clearly display the primitive organization and the essential characteristics of the phyllotaxis of higher plants. For example, in the case of *Fucus spiralis*, illustrated in Figure 5.8, Fibonacci asymmetry is the expression of an oscillatory balancing effect in two dimensions. The diagrammatic representation of this brown alga, obtained from the Horton and Strahler's method [see Sharp (1971), for a presentation of this method] for ordering ramified structures, shows the first terms of the main series.

Some algae show a mechanism of branching by apical bifurcation: when the apex reaches a certain size, it divides into two apices. Recent works on

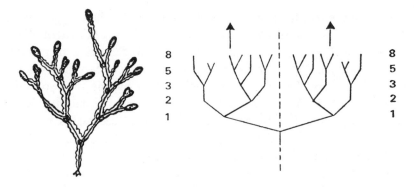

Figure 5.8 The hierarchical organization of *Fucus spiralis* proceeds from a development with unequal dichotomies (overtopping), which can also be identified at the level of vascular phyllotaxis in higher plants (from Jean, 1981c).

phyllotaxis show that the apical size may be an important determinant in the generation of the patterns. McCulloch's algorithm (Research Activity 2b of Section 2.5) suggests a biochemical mechanism for explaining why pinecone scales, for example, split in two at some growth sites but not at others. The botanist Church (1904) frequently insists on the importance of the periodic sequence of ones and twos, $2, 1, 2, 1, 2/2, 1, 2, 1, 2/2, \ldots$ for the explanation of the phenomenon of rising phyllotaxis, analogous to the phenomenon of cellular division. This sequence is found in the consecutive levels of the hierarchy of Figure 5.8.

For Church there is little doubt that the phyllotactic mechanism is an old function of marine vegetation, and the phenomenon of phyllotaxis in flowering plants is just the amplification of phytobenthic factors. Works on vascular phyllotaxis [see Jean (1982), and Research Activity 6 of Section 5.6] show structures identical to that in Figure 5.8; the same is true in the algorithms of reconstruction of certain plants (Jean, 1978b, Section 19).

We are thus led to representing the phyllotactic patterns by multilevel systems showing bifurcations at some points. These are hierarchies with unequal dichotomies, that is, partially ordered systems of interrelated elements—the primordia—interacting in an aggregative fashion. This is in conformity with the theories of the foregoing sections, in particular with the idea that the positioning of a new primordium depends, at most, on the latest two primordia.

The phyllotactic theories generally propose a likely mechanism which tries to reproduce the primordial patterns, those protuberances rising at the surface of the shoot apex. In the approach by hierarchies, arising from a phyletic perception of the phenomenon of phyllotaxis, one does not try to imagine the nature of the actual mechanism, in order to avoid an oversimplification of the tremendous complexity inside the distal zone of the apical meristem, from which the primordia arise. In the cylindrical representation this zone is very simply cut off. The mechanism of phyllotaxis is unknown to investigators but is revealed in hierarchies, the simplifications provided by nature. The initial question becomes, "Why has nature privileged a hierarchy of the type illustrated in Figure 5.8? In a hierarchy we can identify parameters representing the complexity, the stability, and the rhythm, characterizing the growth of plants. A principle of optimum design, based on these parameters, allows us to compare the entropy of these plants. The fundamental hypothesis is that the configuration or pattern is chosen that minimizes the entropy.

5.5.1.2 Definitions. Following are a few definitions related to the axiomatic hypothetico-deductive system of the following section.

A *GROWTH MATRIX* (Jean, 1979a) is a square matrix $C = (c_{ij})$ of order n, whose entries are 0 or 1, where $0 < \Sigma_i c_{ij} \leq 2$ for every j, and whose directed graph is strongly connected. The *LANGUAGE GENERATED* by a growth matrix $C = (c_{ij})$ of order n is the ordered triple (A, P, ω_0) where A is a set with n symbols $a_1, a_2, a_3, \ldots, a_n$, called the alphabet, P is a function defined on A by $P(a_j) = a_1^{c_{1j}} a_2^{c_{2j}} \cdots a_n^{c_{nj}}$ for $j = 1, 2, \ldots, n$, where a_i is in the word $P(a_j)$ if $c_{ij} = 1$ and is not if $c_{ij} = 0$; ω_0 is a word made with the letters of A. If $P(a_j) = a_i a_k$ we say that a_j is a *DOUBLE NODE*; otherwise it is a *SIMPLE NODE*. The *HIERARCHY GENERATED* by C is the sequence $\omega_0, \omega_1, \omega_2, \ldots$ of generations $\omega_t = P(\omega_{t-1})$, $t = 1, 2, 3, \ldots$, the function P behaving like an homomorphism $[P(a_i a_j) = P(a_i)P(a_j)]$. In order to represent graphically a hierarchy we write the word ω_{t-1}, $t = 1, 2, 3, \ldots$, on a horizontal line, and each of its nodes is linked by one or two segments, depending on whether it is a simple or a double node, to the nodes of ω_t, written above ω_{t-1}. A *DUPLICATION* is the set of two segments from a double node in ω_{t-1}. We denote by $S(t)$, $t = 1, 2, 3, \ldots$, the *RELATIVE FREQUENCY OF DUPLICATIONS* in the generations ω_i, $i = 0, 1, 2, \ldots, t$. For example, if ω_0 contains one double node and two simple nodes, $S(1) = 1/3$; if ω_1 contains three double nodes and one simple node, $S(2) = 4/7$; if ω_2 contains one double node and six simple nodes, $S(3) = 5/14$. The (preponderant) period of the *RHYTHM* of a hierarchy is the smallest n such that a growth matrix of order n determines a hierarchy having the same graphic representation as the first (making abstraction of the denomination of the nodes). By definition, a rhythm of period n is induced (a cycle of a mechanism is completed) in n generations $\omega_0, \omega_1, \ldots, \omega_{n-1}$. The *GROWTH FUNCTION* $f(t)$ associated with a hierarchy is the number of symbols in ω_t, $t = 0, 1, 2, 3, \ldots$. We denote by $\langle f(0), f(1) \rangle$ the hierarchies generated by the growth matrix of order 2, $\begin{pmatrix} 1 & 1 \\ 1 & 0 \end{pmatrix}$. The *DEGREE OF COMPLEXITY* of a hierarchy at the tth generation is the number $X(t) = \pi_{k=1}^{t} k^{f(k)}$.

5.5.1.3 Systemic Model (Jean, 1980c). The four following categories of axioms are those of mathematical axiomatic physics. The axioms in C express mutual relations between the variables inside the system. Axiom D expresses the existence of frontiers beyond which other systems exist.

A. Axioms of correspondence.

1. To each type of phyllotactic growth corresponds a hierarchy.
2. To the complexity of a growth corresponds the complexity $X(t)$ of its hierarchy.
3. To the stability of a growth corresponds the relative frequency of duplications $S(t)$ of its hierarchy.
4. To the rhythm of a growth corresponds the rhythm ω of its hierarchy.
5. To the phyllotaxis rising according to the series $\langle H(1), H(2), H(3), \ldots \rangle$, where $H(k + 1) = H(k) + H(k - 1)$, corresponds the set of hierarchies $\langle H(k), H(k + 1) \rangle$, $k = 1, 2, 3, \ldots$.

B. Axiom of configuration. The coordinates of the configuration space are $h, t, T_r,$ and ω: h belongs to the set of hierarchies, t represents a level in the hierarchy, and T_r (in plastochrones) is the number of nodes existing when the rhythm of period ω is established. The dependent variables are $X(t), S(t), P(t)$ the probability of existence of a growth, $W(t)$ the probability of survival of a growth, $I(t)$ the bioentropy of a growth at the tth generation, and $I_b(\omega)$ the bulk entropy of a growth.

C. Axioms of principles inside the system.

1. Principle of existence: $P(t) = 1/X(t)$.
2. Principle of survival: $W(t) = S(t)P(t)$.
3. Principle of entropy: $I(t) = -\log W(t)$.
4. Principle of growth: $I_b(\omega) = \Sigma_{t=1}^{\omega} I(t, h)$.

D. Axiom of principle at the limits of the system. Given T_r a hierarchy is chosen in the following way:

1. Define the subset H of the set of hierarchies, such that for $h \in H$, with period $\omega \geq 2$, $\Sigma_{t=0}^{\omega-1} f(t) = T_r$;
2. Calculate I_b for each $h \in H$;
3. Look for $h \in H$ for which I_b is minimal.

Remarks. Axiom D, a principle of minimization of entropy, is a version of the famous *PRINCIPLE OF OPTIMAL DESIGN*: according to Rosen (1967) to find the optimum solution to a specific problem we must (1) determine the class of all possible solutions, (2) attach to each solution a number representing its cost, and (3) look for the minimal cost in the set of costs; "all the art and the difficulty of the subject lie in finding the appropriate functional cost."

The formula for $I(t)$ allows us to compare the entropy of the phyllotactic systems. This formula is linked to the existing entropy formulas (see Research Activity 8 of Section 5.6) by the factor $\log X(t)$, that is, the entropy of a set of $X(t)$ elements, or the entropy of an event with probability $P(t)$ (see Rényi's definition (1961) for the entropy of an incomplete probability distribution). This formula corresponds to the maximum of the generalized entropy $\Sigma\, p_i \log p_i$, under the minimum of constraints.

Now it is generally admitted that bioentropy must be expressed by a sum of two factors: the formula for $I(t)$ contains two factors. The factor $-\log S(t)$ decreases when $S(t)$ increases, as is the case for the hierarchies corresponding to normal and anomalous phyllotaxis (see Proposition 3 below). It is a factor of negative entropy, "that upon which life feeds" (Schrödinger, 1962), a measure of the vital regulation by which $I(t)$ decreases by a small amount. The capacity to grow implies the ability to reduce entropy.

The numbers of nodes in two consecutive levels of a hierarchy correspond to the secondary numbers of a system (see Research Activity 6 of Section 5.6). When a hierarchy is determined, for all t, the rhythm being established, a divergence angle follows (see, for example, the Theorem of Section 1.3.2 and Exercise 2.14).

The phenomenon of stabilization by duplication is frequent in nature, where an object duplicates when it does not experience sufficient structural stability (see Bruter, 1974a). The value for $X(t)$ is inspired by graphic considerations (Jean, 1978b).

The strong connectivity (irreducibility) of the growth matrices introduces a repetitive factor expressed by the rhythm: all types of nodes return after a finite number of steps, so that a hierarchy has loops. For Church (1920, p. 32) "it is the rhythm that commands the essential mechanism of the phenomenon of phyllotaxis," and the factor inducing the rhythm constitutes the most fundamental problem of physiology. The study of rhythm in biology is the object of this new science called *CHRONOBIOLOGY*. It proceeds from the Darwinian and synthetic (evolution, phylogeny) stream of research in biology. The biological rhythm, a part of the genetic inheritance of the species, occurs as a regular and forseeable variation; one of the first aspects of the temporal structure of an organism is afforded by the preponderant period (see Reinberg, 1977; Gauquelin, 1973).

5.5.2 Consequences of the Model

Proposition 1

Let us consider the hierarchies generated by the companion matrices $C = (c_{ij})$ of order $n = 2, 3, 4, \ldots$, where $c_{11} = c_{12} = \cdots = c_{1n-2} = 0$ and $c_{1n-1} = c_{1n} = 1$, and where $\omega_0 = a_{n-1}$ is the double node of the alphabet $A = \{a_1, a_2, \ldots, a_n\}$. Let us denote by $S_n(t)$ and $f_n(t)$ the corresponding parameters. Then

$$\lim_t S_n(t) > \lim_t S_{n+1}(t);$$

$$S_2(t) > S_n(t) \qquad \text{for } n > 2 \text{ and } t \geq 2;$$

$$\lim_n \lim_t \frac{f_n(t+1)}{f_n(t)} = 1.$$

Proposition 2

A. The hierarchy $\langle n, n + 1 \rangle$ has, for all t, a smaller entropy $I(t)$ than the hierarchy $\langle n + 1, n + 2 \rangle$ for $n = 1, 2, 3, \ldots$.

B. The hierarchy $\langle 1, 2 \rangle$ has a smaller bulk entropy than the hierarchies in Proposition 1 and the hierarchy that has double nodes only.

C. Among all the hierarchies having at least one double node in ω_0, the values for $I(1)$ and $I(2)$ are minimal for the hierarchy $\langle 1, 2 \rangle$ and for the hierarchy in Proposition 1 with $n = 3$.

D. If the hierarchies $\langle f(0), f(1) \rangle$ are set in order by (strictly) increasing values of I_b, then we have consecutively $\langle 1, 2 \rangle, \langle 2, 3 \rangle, 2\langle 1, 2 \rangle,$ $\langle 3, 4 \rangle, \langle 3, 5 \rangle, 3\langle 1, 2 \rangle, 2\langle 2, 3 \rangle, \langle 4, 5 \rangle, \langle 4, 7 \rangle, 4\langle 1, 2 \rangle, \langle 5, 7 \rangle, \langle 5, 6 \rangle, \langle 5, 8 \rangle,$ $\langle 6, 7 \rangle, \dots$.

E. Among the hierarchies mentioned in B and in D, $\langle 1, 2 \rangle$ has the maximal stability for all t.

F. If $T_r = 3, 5, 6, 7, 8, 9, 10, 11, 12, 13,$ and 14, I_b is minimal (axiom D) for $\langle 1, 2 \rangle, \langle 2, 3 \rangle, 2\langle 1, 2 \rangle, \langle 3, 4 \rangle, \langle 3, 5 \rangle, 3\langle 1, 2 \rangle, 2\langle 2, 3 \rangle, \langle 4, 7 \rangle, 4\langle 1, 2 \rangle,$ $\langle 5, 8 \rangle$, and $\langle 5, 9 \rangle$, respectively (for $T_r = 9$ we have $3\langle 1, 2 \rangle$ instead of $\langle 4, 5 \rangle$ or of a hierarchy which has double nodes only; for $T_r = 11$ we have $\langle 4, 7 \rangle$, not $\langle 5, 6 \rangle$; for $T_r = 12$, we have $4\langle 1, 2 \rangle$, not $\langle 5, 7 \rangle$; and for $T_r = 13$, $\langle 5, 8 \rangle$ arises instead of $\langle 6, 7 \rangle$).

Proposition 3

Denoting by $S_n(t)$ and $S_n^*(t)$ the relative frequency of duplications in the hierarchies $\langle n, n + 1 \rangle$ and $\langle 2n + 1, 3n + 1 \rangle$, respectively, we have

$$S_n(t) = \frac{F(t + 1)n + F(t) - n}{F(t + 2)n + F(t + 1) - (n + 1)}, \qquad n = 1, 2, 3, \dots,$$

$$S_n^*(t) = \frac{F(t + 3)n + F(t + 1) - (2n + 1)}{F(t + 4)n + F(t + 2) - (3n + 1)}, \qquad n = 2, 3, 4, \dots,$$

$$S_n(t) > S_{n+1}(t), S_n^*(t) < S_{n+1}^*(t), \lim_n S_n^*(t) = S_2(t),$$

$$S_1(t) > S_n(t), S_1(t) > S_2(t) > S_n^*(t), \qquad n = 2, 3, 4, \dots,$$

$$\lim_t S_n(t) = \lim_t S_n^*(t) = \phi^{-1},$$

S_n and S_n^* are increasing functions of t for $n = 2, 3, 4, \dots$ and S_1 is a decreasing function of t (strictly for $t = 3, 4, 5, \dots$).

Proposition 4

For all t, the minimal value $I^*(t)$ of $I(t)$, such that ω_0 contains at least one double node, is

$$I^*(t) = \log(2t - 1) + 2 \sum_{k=1}^{t} \log k.$$

The hierarchy in Proposition 1, n fixed, has, among all the hierarchies having a period of rhythm smaller than or equal to n, minimal entropy $I(t)$ for $t = 1, 2, \ldots, n$, and $I(t) = I^*(t)$ for $t = 1, 2, \ldots, n - 1$.

Proof. When ω_0 contains only one double node, $I(2)$ has the same value for two simple nodes, or for one simple and one double node, in ω_1. In the last case, if we suppose that ω_i, $i = 2, 3, \ldots, t - 1$, contain three nodes only, then $X(t) = (t!)^3$ is the minimal value that $X(t)$ can take. If we suppose that $S(t) = 1$, then $3 \log t!$ minimizes $I(t)$. But this value is larger than $I^*(t)$ for $t = 3, 4, 5, \ldots$.

If there are at least two nodes in ω_0, then $I(1) \geq 0$, and for $t = 3, 4, 5, \ldots$, supposing that $S(t) = 1$ and that $X(t)$ has the smallest possible value, that is, $(t!)^3$, $I(t) = 3 \log t!$, a number greater than $I^*(t)$. For $t = 2$ it can be verified that $I^*(2)$ is the smallest possible value.

The hierarchy determined by the matrix of order n of Proposition 1 is the one for which $I(t) = I^*(t)$ for $t = 1, 2, 3, \ldots, n - 1$. If ω_{n-1} contained two simple nodes only, we would have $I(n) = I^*(n)$, but the period of the rhythm would be larger than n. The minimum is obtained from the case where there is one double node and one simple node in ω_{n-1}. ∎

Proposition 5

Given an increasing series of integers $\langle H(1), H(2), H(3), \ldots \rangle$, such that $H(k + 1) = H(k) + H(k - 1)$, and $T_r = H(k + 3)$, such that $H(k + 2) \leq 2H(k + 1)$, then the hierarchy

$$\left\langle H(k + 1) + \left[\frac{-H(k - 1)}{3} \right], H(k + 2) - \left[\frac{-H(k - 1)}{3} \right] \right\rangle$$

where $[\cdot]$ means the integral part of, has, among the hierarchies for which

$\omega = 2$, minimal $I(1)$, minimal $I(2)$, and minimal I_b given by

$$I_b = \log\left(\frac{H(k+1)+m}{H(k)-2m}\right)\left(\frac{H(k+3)}{H(k+2)-m}\right)2^{H(k+3)},$$

where m is an integer such that

$$\frac{-H(k-1)}{3} \leq m < \frac{H(k)}{2}.$$

Proof. Whatever the value of T_r may be, when $\omega = 2$, the minimal bulk entropy is given by one of the hierarchies generated by $C = \begin{pmatrix} 1 & 1 \\ 1 & 0 \end{pmatrix}$, given that the hierarchies generated by $\begin{pmatrix} 1 & 1 \\ 1 & 1 \end{pmatrix}$ are such that $f(0) + f(1)$ is a multiple of 3, that is, $3p$, and that $p\langle 1, 2\rangle$ has a smaller bulk entropy. The only other growth matrix for which $\omega = 2$ is $\begin{pmatrix} 0 & 1 \\ 1 & 0 \end{pmatrix}$. In that case $I(t) = \infty$ for all t.

Thus the hierarchy for which $T_r = H(k+3)$, $\omega = 2$, and I_b is minimal, is such that $3x + 2y = H(k+3)$, where x and y are respectively the numbers of double nodes and simple nodes in ω_0. This equation gives $x = H(k) - 2m > 0$, $y = H(k-1) + 3m \geq 0$, and m is as in the statement of the proposition. We thus have the hierarchies $\langle H(k+1) + m, H(k+2) - m\rangle$, for which

$$I_b = I(1) + I(2) = \log\left(\frac{x+y}{x}\right)\left(\frac{3x+2y}{2x+y}\right)2^{3x+2y},$$

that is the value of I_b given in the statement. As a function of m, I_b is increasing in the above interval, so that for $m = [-H(k-1)/3]$, I_b is minimal. This value of m also minimizes $I(1)$ and $I(2)$. ■

Proposition 6

The bulk entropy is smaller for the hierarchy $\langle H(k+1), H(k+2)\rangle$ having at least one double node in ω_0, than for any other hierarchy with period $\omega > 2$ having $H(k+3) \geq 3$ nodes in the first ω generations (including ω_0).

Proof. For the hierarchy $\langle H(k+1), H(k+2)\rangle$, $I_b = I_b'$ is smaller or equal to $\log(H(k+3)/H(k))2^{H(k+3)}$, a value approximately equal to

$\log \phi^3 2^{H(k+3)}$. Consider a hierarchy with period $\omega > 2$, having $H(k+3)$ nodes in the first ω generations. If we suppose that for this hierarchy $S(t) = 1$ for all t, and $X(t) = (t!)^{H(k+3)/\omega}$ for $t = 2, 3, \ldots, \omega$, then the bulk entropy of the hierarchy will be minimized by the value $E = \log \pi_{t=2}^{\omega} X(t)$. It is easy to show that for $\omega \geq 4$, $I_b' < E$, and the same for $\omega = 3$ and $H(k+3) \geq 8$. Finally, for $3 \leq H(k+3) \leq 7$ and $\omega = 3$, the question can be solved by working directly on the possible hierarchies. ■

Proposition 7

Given an increasing series of integers $\langle H(1), H(2), H(3), \ldots \rangle$ such that $H(k+1) = H(k) + H(k-1)$, and $T_r = H(k+3)$ such that $H(k+2) \leq 2H(k+1)$, then the hierarchy $\langle H(k+1) + [-H(k-1)/3], H(k+2) - [-H(k-1)/3] \rangle$ has, among all the generated hierarchies, minimal values for I_b, $I(1)$, and $I(2)$.

Proof.   This is a consequence of Propositions 5 and 6. ■

Theorem.   Let $\langle H(1), H(2), H(3), \ldots \rangle$ be an increasing series of integers where $H(k+1) = H(k) + H(k-1)$ and $H(k+2) \leq 2H(k+1)$; then we have the rising phyllotaxis $\langle H(1), H(2), H(3), \ldots \rangle$ if and only if there is a $k \in \{0, 1, 2, 3, \ldots\}$ such that for $T_r = H(k+3)$ we have $H(k-1) = 0, \pm 1,$ or ± 2.

Proof.   By axiom A5 and axiom D, there is a k such that for $T_r = H(k+3)$, $\langle H(k+1), H(k+2) \rangle$ has a minimal I_b. By Proposition 7, $[-H(k-1)/3] = 0$, that is, $H(k-1) = 0, \pm 1,$ or ± 2.

Conversely, suppose that k is such that $T_r = H(k+3)$, $H(k-1) = 0, \pm 1,$ or ± 2. Then by Proposition 7, the hierarchy $\langle H(k+1), H(k+2) \rangle$ has a minimal I_b. This is the hierarchy chosen by axiom D, and by axiom A5 this hierarchy represents the rising phyllotaxis $\langle H(1), H(2), H(3), \ldots \rangle$. ■

Corollary 1.   The patterns obtained for a few values of T_r are as follows:

$$\langle 1, 2, 3, 5, 8, 13, \ldots \rangle \Leftrightarrow T_r = 3, 5, 8, 13;$$

$$2\langle 1, 2, 3, 5, 8, 13, \ldots \rangle \Leftrightarrow T_r = 6, 10, 16;$$

$$\langle 1, 3, 4, 7, 11, 18, \ldots \rangle \Leftrightarrow T_r = 7, 11;$$

$$3\langle 1,2,3,5,8,13,\ldots\rangle \Leftrightarrow T_r = 9;$$

$$4\langle 1,2,3,5,8,13,\ldots\rangle \Leftrightarrow T_r = 12;$$

$$\langle 1,4,5,9,14,\ldots\rangle \Leftrightarrow T_r = 14;$$

$$5\langle 1,2,3,5,8,13,\ldots\rangle \Leftrightarrow T_r = 15;$$

$$\langle 1,5,6,11,17,\ldots\rangle \Leftrightarrow T_r = 17;$$

$$6\langle 1,2,3,5,8,13,\ldots\rangle \Leftrightarrow T_r = 18;$$

$$\langle 2,5,7,12,19,\ldots\rangle \Leftrightarrow T_r = 19;$$

$$\langle 1,6,7,13,20,\ldots\rangle \Leftrightarrow T_r = 20;$$

$$2\langle 1,3,4,7,11,18,29,\ldots\rangle \Leftrightarrow T_r = 22;$$

$$\langle 2,7,9,16,\ldots\rangle \Leftrightarrow T_r = 25;$$

$$2\langle 1,4,5,9,14,23,\ldots\rangle \Leftrightarrow T_r = 28;$$

$$\langle 2,9,11,20,\ldots\rangle \Leftrightarrow T_r = 31;$$

$$2\langle 1,5,6,11,17,\ldots\rangle \Leftrightarrow T_r = 34.$$

Corollary 2. We have the systems $\langle 1,2\rangle$ or $\langle 2,3\rangle$ if and only if $T_r < 6$.

Corollary 3. A necessary and sufficient condition to have the phyllotaxis $\langle 1,2,3,5,8,13,\ldots\rangle$ is that $T_r < 6$, $T_r = 8$, or $T_r = 13$.

Corollary 4. The only types of phyllotaxis that can possibly exist are given by the sequences

$$J\langle 1,2,3,5,8,13,\ldots\rangle, J \geq 1;$$

$$J\langle 1,t,t+1,2t+1,3t+2,\ldots\rangle, J = 1 \text{ or } 2 \text{ and } t \geq 3;$$

$$\langle 2,2t+1,2t+3,4t+4,6t+7,\ldots\rangle, t \geq 2.$$

For all T_r (including $T_r = 2$ and 4, which would correspond to the distichous and decussate patterns) one and only one of these systems arises. The corresponding divergence angles are equal to

$$\left[J\!\left(t + \phi^{-1}\right) \right],^{-1} \qquad \left[2 + \left(t + \phi^{-1}\right)^{-1} \right]^{-1}.$$

Proof. T_r can have the form $3p$, $p \geq 1$; $3p + 1$, $p \geq 2$; or $3p + 2$, $p \geq 1$. The corresponding hierarchies, satisfying the theorem are $\langle p, 2p \rangle, \langle p + 1, 2p \rangle, \langle p + 1, 2p + 1 \rangle$. They represent the phyllotaxis $\langle p, 2p, 3p, \ldots \rangle$, $p \geq 1$; $\langle p + 1, 2p, 3p + 1, \ldots \rangle$, $p \geq 2$; $\langle p + 1, 2p + 1, 3p + 2, \ldots \rangle$, $p \geq 1$. Rearranging these sequences and eliminating the duplications (such as $p = 2$ in the first series and $p = 3$ in the second), we have the specified sequences. For the second value of the divergence see Exercise 2.14 in Appendix 2. ∎

5.5.3   Interpretation, and By-Products of the Model (Jean, 1980c)

The types of phyllotaxis in Corollary 1 are, in order presented, more and more unlikely: in each row, downward, the number of T_r's decreases and the value of T_r increases. The fact that the series $\langle 1, 2, 3, 5, 8, 13, \ldots \rangle$ is frequently represented first in the list, and that the hierarchy $\langle 1, 2 \rangle$ has a maximal stability and a minimal bulk entropy can explain the omnipresence of this type of phyllotaxis in nature. For $T_r \geq 6$ it is still possible, according to Corollary 3, to obtain normal phyllotaxis with $t = 2$ (Expression 2.1): Cutter (1956) reports that on *Dryopteris aristata* this phyllotaxis settles only when there are at least six primordia, and Majumdar (1948) points out that for the sunflower the stable pattern of the leaves is reached only after sufficient growth of the plant, showing first the opposed and then the decussate systems.

According to Corollary 1 the phyllotaxis $\langle 1, 2, 3, 5, 8, 13, \ldots \rangle$ would be more frequent than the phyllotaxis $\langle 1, 3, 4, 7, 11, 18, \ldots \rangle$, which would be more frequent than $\langle 1, 4, 5, 9, 14, 23, \ldots \rangle$, itself more frequent than $\langle 2, 5, 7, 12, 19, \ldots \rangle$: this corresponds to reality. From the same corollary, the bijugate system $2\langle 1, 2, 3, 5, 8, 13, \ldots \rangle$ would be more frequent than the system $\langle 1, 3, 4, 7, 11, 18, 29, \ldots \rangle$: this does not seem to have been verified, although the observations made by Fujita (1938), reported by Williams (1975, p. 31), on more than 500 species of flowering plants, seem to confirm this conclusion. The plant showing a more usual type of phyllotaxis would be more suited to the following process: fixing solar energy and transforming it for its needs, and thereby synthesizing material that can be used by other organisms. It would perhaps be possible to make a comparative study of the wavelengths absorbed by the chloroplasts and carotenoids of each species, of quanta of energy fixed by photophosphorylation, of the potential energy stored in the cytoplasm in the form of ATP, etc.

According to Fujita (1937), the sequences encountered in phyllotaxis are, a priori, the following:

$$\langle 1, t, t + 1, 2t + 1, 3t + 2, \ldots \rangle, \qquad t \geq 1;$$

$$\langle p, ap + 1, (1 + a)p + 1, (1 + 2a)p + 2, \ldots \rangle, \qquad p \geq 2, a \geq 2;$$

$$\langle q, bq - 1, (1 + b)q - 1, (1 + 2b)q - 2, \ldots \rangle, \qquad q \geq 3, b \geq 3.$$

In the case of whorled patterns, the terms of these sequences must be multiplied by the number of elements in the whorl (see Research Activity 5c in Section 2.5). For Fujita's second sequence, the cases $p = 2$ and $a = 2, 3, 4$ have frequently been observed (see Fujita, 1937); they correspond to the third sequence in Corollary 4 with $t = 2, 3, 4$. Now the cases having a probability of existing are those where $p = 2$ and $a \geq 2$, so that it is possible that the cases where $a = 5, 6, 7, \ldots$ will be found in nature. The theory says that cases where $p \geq 3$ will never be found. For the second sequence in Corollary 4, the cases where $J = 1$ and $t = 1, 2, 3, \ldots, 17, 18$ are known to exist in nature; some multijugate systems ($J > 1$) have also been observed. Quoting Braun and Hofmeister, Fujita asserts that the cases $q = 3$ and $b = 3, 5$, giving the particular sequences $\langle 3, 8, 11, 19, 30, \ldots \rangle$ and $\langle 3, 14, 17, 31, \ldots \rangle$, would be found in *Grimmia leucophaea* (7/19) and *Monstera deliciosa*. The patterns on these plants must be reinterpreted (the types of phyllotaxis for which I_b is minimal when $T_r = 19$, 30, or 31, respectively, belong to the sequences $\langle 2, 5, 7, 12, 19, \ldots \rangle$, $10\langle 1, 2, 3, 5, 8, \ldots \rangle$, and $\langle 2, 9, 11, 20, 31, \ldots \rangle$; 8/19 is a convergent of the divergence corresponding to the first sequence).

This approach to phyllotaxis delivered results on the asymptotic behavior of the ratios $f(t + 1)/f(t)$, C^t/r^t, and $f(t)/r^t$, where r is the spectral radius of C and f the associated growth function (see Section 5.5.1.2). These results make possible the analysis of the growth of organisms generated by PDOL systems, where C is an irreducible matrix, and inform as to the possibility of

Table 5.6   Values of $\lim_t f(t - m) / f(t - n)$ obtained from the characteristic equation $\lambda^{m+n-1} - \lambda^{m-1} - \lambda^{n-1} = 0$ of an irreducible matrix whose growth function is f (from Jean, 1981a)

				n				
m	1	2	3	4	5	6	7	8
1	1	ϕ	2.1478	2.6296	3.0795	3.5063	3.9150	4.3093
2	$1/\phi$	1	1.3247	ϕ	1.8905	2.1478	2.3935	2.6296
3	0.4655	0.7548	1	1.2207	1.4252	ϕ	1.8016	1.9778
4	0.3802	$1/\phi$	0.8191	1	1.1673	1.3247	1.4744	ϕ
5	0.3247	0.5289	0.7016	0.8566	1	1.1347	1.2627	1.3854
6	0.2851	0.4655	$1/\phi$	0.7548	0.8812	1	1.1127	1.2207
7	0.2554	0.4177	0.5550	0.6782	0.7919	0.8986	1	1.0969
8	0.2320	0.3802	0.5056	$1/\phi$	0.7217	0.8191	0.9115	1

approximating $f(t)$ by kr^t, and on the value of $f(t + h)/f(t)$. These are relevant results in population biology where C is a Leslie matrix. They provide relations between the theory of OL systems and Perron-Frobenius' spectral theory, that is, a theory of languages and a field of functional analysis (see Jean, 1981a).

Table 5.6, meant to aid botanists, is an application of these results. It concerns in particular the development of lineages of cells, of filamentous algae, or of ramified structures, by means of L systems. It gives the value of $\lim_t f(t - m)/f(t - n) = r^{n - m}$ for current values of m and n. Its entries can be compared to the data observable on plants, and they allow us to establish the agreement between the predicted number of cells and those that can be enumerated. The three limits in the preceding paragraph do not exist when $2m = n$, but do exist when m and n are relatively prime numbers, as is the case for the blue-green alga *Anabaena catenula* ($m = 4$, $n = 5$), the red alga *Chaetomorpha linum* ($m = 3$, $n = 5$), the red alga *Callithamnion roseum* and the moss *Protonema* ($m = 1$, $n = 5$), and the leaves of the fern *Athyrium filix-femina* ($m = 1, n = 3$).

5.6   RESEARCH ACTIVITIES

1. *Relation between three leaves near the origin.* The result in Exercise 2.34 is fundamental in Adler's model (Section 5.2) and in Mitchison's report

(1977). The latter supposes that the leaves are tangent circles having the same diameter. What he calls the *CONTACT CIRCLE* of a leaf is the locus of the possible positions of the centers of its two contacts; the contacts of a leaf are its nearest neighbors, and the contact circle of leaf 0 is fixed. Assuming that the centers of the leaves m and n are on the contact circle of leaf 0, Mitchison declares that the next point of the cylindrical lattice to meet the contact circle can only be the leaf $m + n$. This is essentially Exercise 2.34. Then by virtue of his *2/1-PHYLLOTAXIS PRINCIPLE*, the phyllotaxis rises along the consecutive terms of the main series. According to this principle, at the very beginning of growth, leaf 0 and leaf 1 are on opposite sides of the stem, meeting at two points (distichous pattern). With the increase of the apex diameter, brought forth by the growth of the plant, the decrease of the slope of the genetic spiral (the rise r decreases), and the birth of a new leaf, the two sides of leaf 1 (formerly 0) depart from the two points of contact with leaf 2 (formerly 1), thus allowing the new leaf 0 to touch leaf 2 (the numbering changes each time a new leaf 0 arises). When this happens, leaf 0 occupies one of two possible positions, on each side of leaf 2, generating a clockwise or counterclockwise spiral, and a system whose phyllotaxis is 2/1. This initial pattern on the contact circle is assumed to be the first cause of Fibonacci phyllotaxis; it would be determined, according to Mitchison, by an inhibitor secreted by the apical dome and the existing leaves. This seems to produce Fibonacci phyllotaxis as an inevitable consequence, and to make all the other cases impossible! Whatever it may be, following is a summary made from another proof by Adler (personal communication, November 12, 1977), of the result mentioned earlier, and according to which for $T > T_c$, three neighbors of the origin are related by the recurrence relation defining the main series. Develop it completely.

Proposition

Let $v = d(0, m) = d(0, n)$ be the minimal distance between the points in the lattice, $w = d(0, m + n)$, and $x > m$, $x > n$, $x \neq n + m$, where m and n are relatively prime numbers. Then $d(0, x) > w$ or $d(0, x) > v$.

Proof. Let θ be the angle at 0 determined by the points 0, m, and n. Because $v \leq w$, it can be shown that $\pi/3 \leq \theta \leq 2\pi/3$. Given that $x = am + bn$, where a and b are integers, if a or $b = 0$ then $d(0, x) > v$.

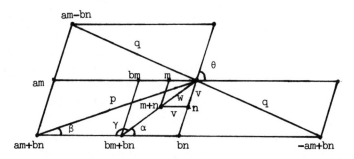

Figure 5.9 Leaf $m + n$ is the next one after the leaves m and n to come on the contact circle of leaf 0.

Supposing, without loss of generality, that $a \geq b > 0$, three cases can occur: (1) $x = am + bn$, (2) $x = am - bn$, and (3) $x = -am + bn$. Putting $d(0, am + bn) = p$ and $d(am, bn) = q$, for the first case it is sufficient to show that $p > w$, and for the other two cases that $q > w$ or that $q > v$. If $a = b$ it is easily shown that $p > w$ and that $q > v$. Assuming that $a > b$, then $d(0, bm + bn) = bw \geq w$. Since that $p > bw$, the first case is settled: in Figure 5.9 one sees that $\alpha = \theta/2$, and consequently that $\pi/6 \leq \alpha \leq \pi/3$ and $\gamma > \alpha > \beta$. In the last two cases we have $q^2 = v^2(a^2 + b^2 - 2ab\cos\theta)$, and $w^2 = 2v^2(1 + \cos\theta)$. If $\theta \leq \pi/2$, $v^2(a^2 + b^2 - ab) \leq q^2 \leq v^2(a^2 + b^2)$ and $2v^2 \leq d^2 \leq 3v^2$: putting $a = b + c$ we have $q^2 \geq v^2(b^2 + c^2 + bc) \geq 3v^2 \geq d^2$, and $q > v$, $q > w$ (except for $\theta = \pi/3$). If $\theta > \pi/2$, $v^2(a^2 + b^2) \leq q^2 \leq v^2(a^2 + b^2 + ab)$, and $v^2 \leq w^2 < 2v^2$: $q^2 \geq v^2(a^2 + b^2) > 2v^2 > w^2$, and $q > v$, $q > w$. ∎

2. *Spiral formations and contact pressures.*

   (a) Exhibit Church's criticism of Schwendener's theory, and present Church's theory according to which the spiral formations are primitive, the other formations being phylogenetically related to the former (see Church, 1904, pp. 46, 51, 236–266, and Jean 1978b, pp. 100–103; Beck et al., 1982, p. 813).

   (b) Set out Roberts' algorithm (1977) for generating by contact pressures the semidecussate and other such patterns from a distortion of spiral formations.

3. *Roberts' variants of Adler's model.*

   (a) Adler (1975, p. 444; 1977a, p. 48) declares that as a consequence of Proposition 4 in Section 5.2.3, a complete model of phyllotaxis is obtained when the contact pressure model is combined with the weakest form of the model in Section 5.3, guaranteeing that consecutive primordia will generate a fundamental spiral. Set out Roberts' reasoning (1978, p. 219) meant to invalidate this assertion.

(b) Present the five variants of the contact pressure model, proposed by Roberts (1978) in order to improve Adler's model.

4. *Likelihood of the diffusion theories.* In a short article, Richter and Schranner (1978), realizing that the emergence of the main series in phyllotaxis is simply the strict consequence of the appearance of ϕ^{-2} along the genetic spiral, propose an argument according to which all the types of phyllotaxis can be obtained from two main characteristics of the inhibiting interaction, namely, its lifetime and spatial range. Make a detailed report of the content of this article (see Research Activity 12).

5. *Variable phyllotaxis.* In many seedlings, spiral phyllotaxis derives from the decussate pattern via the distichous pattern, and the rising phyllotaxis proceeds from an increase of the apex (see Camefort, 1956; Richards, 1951; Gifford and Tepper, 1962). These ontogenetic changes in phyllotaxis are an aspect of heteroblastic development characterized by a well-marked difference between the juvenile and the mature states. Record cases of such natural changes (see Allsopp, 1965).

Describe the various types of phyllotaxis observed on *Bryophyllum tubiflorum* (see Gomez-Campo, 1974), and present Hellendoorn and Lindenmayer's simulation (1974) of this variable phyllotaxis, making use of Expression 4.31.

6. *Vascular phyllotaxis and hierarchical representation.* The question of vascular differentiation is one of very great interest in the study of plant morphogenesis. In particular, the concept of foliar trace is very useful when dealing with phyllotaxis and other related phenomena. Figure 5.10 enables us to relate the phyllotaxis of a *Sequoia* shoot to the links between its foliar traces. The traces are connected by the 13-parastichies of a system where the (5, 3) parastichy pair is conspicuous, and between two leaves on a bundle there are five circuits around the tip of the apex ($d \simeq 5/13$). Sterling (1945) suggested that the bundle arrangements can be best described by taking into account the direction of the ontogenetic spiral, the angle of divergence and the sympodial number. In Figure 5.10 the sympodial number is 13 (there are 13 discrete bundles from which leaf traces diverge). In conifers with helical phyllotaxis, the

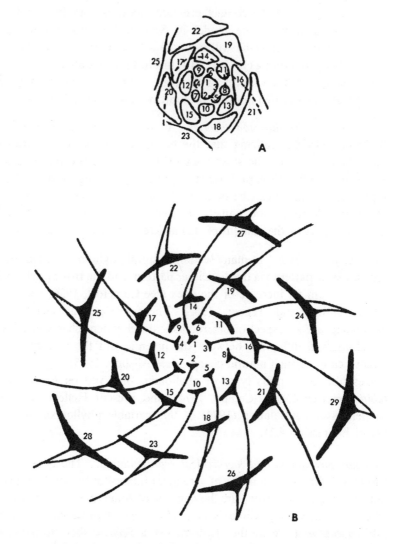

Figure 5.10 Relation between the phyllotaxis (A) and the foliar traces (B) of a shoot of *Sequoia sempervirens* (from Esau, 1954, 1965b). The triangular areas on the apex viewed from above (in B) represent the insertions of the primordia of the transverse section (in A).

phyllotactic fractions generally belong to Schimper and Braun's series (2/5, 3/8, 5/13, 8/21) and the numbers of sympodia to the Fibonacci sequence (5, 8, 13, 21) (see Namboodiri and Beck, 1968).

Works such as those of Esau (1943) on *Linum*, of Sterling (1945) on *Sequoia*, of Gunckel and Wetmore (1946) on *Ginkgo*, and of Tucker (1962) on *Michelia* bring to light the fact that the primordial pattern is determined, in some measure, by the vascular structure of the shoot. Impressed by the close relation between the procambium, or incipient vascular tissue, and the pattern of leaves, some assert that the acropetal development of procambial bundles projected into the region of a primordium before its birth must be, if not a control factor of the phyllotaxis, at least a corollary of Snow's hypothesis (Section 5.3.1). These observations enable us to understand why a protuberance (primordium) arises on the surface of the apical meristem, whatever it may become afterwards. Snow and Snow (1948) are opposed to this theory of foliar induction, and experiments such as those of Wardlaw (1949c) on *Dryopteris* seem to show that the primordium determines the procambium. Whether or not the prevascular bundles and the primordia arise together, their position would be at least determined by the same factor (Philipson, 1949). Esau (1954) proposes a detailed discussion of this problem. For Steward (1968) it is clear that the problem of phyllotaxis concerns the vascular system, and for Larson (1977), working on *Populus*, "the vascular organization dictates the order of phyllotaxis" (see also Larson, 1983; Namboodiri and Beck, 1968).

The hierarchical nature of the primordial patterns in phyllotaxis can be made conspicuous, with works on vascular phyllotaxis such as those of Larson (1977) on *Populus*, Jensen (1968) on *Kalanchoe*, Vieth (1965) and Bolle (1939) on *Cephalaria*, Girolami (1953) on *Linum*, Gunckel and Wetmore (1946) on *Ginkgo biloba*, and Nägeli (1858) on *Iberis amara*. Make an illustrated presentation of these works (see Jean, 1982). Dormer (1972, Chapter 6) and Esau (1943; 1965b, chapter 2) are good introductions to the prominent concepts related to the structure of the vascular system. Consult Clowes (1961) and Wardlaw (1965a). *ZIMMERMANN'S TELOME THEORY* (1953) (see Stewart, 1964) asserts that vascular plants have evolved from a primitive branching system and proposes algal ancestors of early vascular plants; Hofmeister (1868) formulated a similar idea. The principle of overtopping (see Fig. 5.8) is an essential part of the telome theory which has been fruitfully applied to conifers by Florin (1951). According to him the overtopping process means that the leaves on the shoots will primarily be arranged alternately.

7. *What a diffusion theory can, and cannot do.* The Epilogue presents the Assumptions R1, R2, and C3. Expose Adler's counter-example (1974, p. 28), showing that Richards' diffusion theory (R1 and R2) does not necessarily imply ϕ^{-2} (C3).

8. *The notions of entropy.* Schrödinger (1962) introduced the concept of entropy into the life sciences by developing a literal interpretation of the famous formula $E = k \log D$, where D is a measure of the disorder and k is Boltzmann's constant. In the base 2 this formula becomes $E = \log D$. Wiener pointed out the need to extend the notion of physical entropy when he declared that "Information is negative entropy." For Papentin (1980), "There is an irreducible subjectivity associated with the problem of complexity and the related problem of order." He defined the information content I of an "event" by the amount of "choice" inherent in it: $I = \log N$ where N is the number of possibilities available. The thermodynamics of nonequilibrium phenomena is based on the entropy balance $dS = d_e S + d_i S$, where $d_e S$ is linked to the exchanges of energy and matter with the external environment, and $d_i S > 0$ is the entropy production due to the irreversible reactions (such as diffusion) inside the system (see Glansdorff and Prigogine, 1971).

Present various concepts of entropy, such as Rashevsky and Trucco's structural information content (Trucco, 1956), Shannon and Weaver's formula (1967), Mowshowitz's chromatic information content (1967), and the entropy of a partition, in order to demonstrate some of their mutual relations, their similarities, and the need for defining, particularly for the open systems, entropy measures. Brillouin (1949) can be consulted for an intuitive approach to the subject. Steward (1968) presents an analysis of entropy as it is related to growth. Kuich (1970) defines the entropy H associated with a language as the logarithm of the inverse of the radius of convergence of the function $f(z) = \sum_{n=1}^{\infty} u(n) z^n$, where $u(n)$ is the number of distinct words of length n in the language. Guiaşu (1977) presents the notions of weighted and conditional entropy, of ϵ-entropy, of entropy of order α, of absolute S-entropy, and so forth in the framework of measure theory. Haynes (1980) shows that no universally accepted definition of the notion entropy exists; he records and comments on those that do exist. Jean (1980a) presents a study related to the notion of entropy in Section 5.5. See also von Bertalanffy (1973), Brillouin (1959), Khintchine (1957), Lehninger (1969), Pilet (1967), and Demetrius (1981).

9. *Snow's theory versus Schoute-Richards' theory.* Analyze the microsurgical morphogenetic experiments meant to verify Snow's or Schoute-Richards' hypotheses, setting out the arguments in favor of one or the other. Snow and Snow (1952), Clowes (1961), Wardlaw (1965a, 1968a), and the few references given in Sections 4.1 and 5.3.1 can be consulted.

10. *Simulation of Adler's model for the sunflower.* Present Ridley's computer simulation (1982b) of the growth by contact pressure of a capitulum, in order to verify Adler's model for the case where phyllotaxis is expressed by high secondary numbers.

11. *The systemic approach with projective geometry.* Considering the living organism as a whole which is greater than the sum of its parts (in contrast to the reductionist, for whom the organism is simply the aggregate of its parts) Edwards (1982) develops, in his charming little book, Steiner's idea of using projective geometry to enlighten the genesis of form and to illustrate the way in which it can survive changes.

(a) Show how, by a very elementary linear process, Klein and Lie's continuous path curves can be generated, defining the notions of growth measure, of multiplier, of invariant triangle and of invariant tetrahedron. What do the relations $x^a y^b z^c = k$ and $\alpha\beta\gamma = 1$ represent? Notice that the case where the invariant triangle has two of its vertices projected to infinity gives the equation $y = kx^a$, that is, the allometric formula met in many sectors of biology, which establishes a correlation of growth between the parameters x and y of an organism [see Section 3.3.5, Jean (1983e), Huxley (1972)].

(b) Illustrate the notions of elliptical metric, semiimaginary tetrahedron, and pivot transformation, generating families of logarithmic spirals.

(c) Show how Edwards uses these notions to initiate the spirality observed in plant formations such as cones, buds, inflorescences, and flowers, by means of measures taken on enlarged photographs of these plants.

12. *The set Φ_G, locus of the minimal concentration of the morphogen.* Taking into consideration only the divergence $\Delta\theta$ in a regular lattice, as in Thornley's model (Section 5.4.1), Marzec and Kappraff (1983) define a class of concentration fields of a morphogen produced by the leaves i located at θ_i by $C_{\Delta\theta}(\theta) = \Sigma\lambda^i f(|\theta - \theta_i|)$, where $0 < \lambda < 1$. The contribution of leaf i to the concentration function C decreases with time, and the only restrictions on the diffusion function f are those that guarantee that the value of $C(\theta)$ is influenced most strongly by the leaves near θ. Show that there exists a large class of functions f for which the absolute minimum of the concentration C arises when $\Delta\theta$ belongs to Φ_G, the set studied in Research Activity 7 of Chapter 2.

6 EPILOGUE: THE STATE OF THE ART

It is because of investigations into phyllotaxis that new ideas have brought about the considerable progress achieved in the last two decades in our knowledge of plant apices. (Loiseau, 1969)

6.1 PRESENTATION

At the beginning of the century, Church (1904) pointed out the variety of hypotheses put forward to explain phyllotaxis, with none being confirmed or invalidated. With the evolution of experimental research, three main theories claimed attention in the 1940s, with various amounts of factual support. There were the first available space or space filling theory, implicit in van Iterson's work (1907) and developed by Snow and Snow (1931, 1962), and the diffusion theory of Richards (1948, 1951), originating in Schoute's work (1913) and favored by Wardlaw (1952) after his experiments on *Dryopteris* and by Schwabe (1971). The inhibitor theory then looked more attractive, more physiological, and less formal than Snow's theory. The experiments of the latter on *Lupinus albus* (1955, 1962) sustained his hypothesis. The third theory is Plantefol's foliar helices and contiguity theory (1948, 1950), omnipresent in the French literature on morphology. This theory, against which much has been written (see, for example, Snow, 1948, 1949, 1955; Richards, 1951, pp. 513–514; or Sinnott, 1960, p. 157), accounts, according to Philipson (1949), for many facts of the apical development. Wardlaw (1965a) made an extensive critical analysis of it. According to Cutter (1959, 1965), "It is not supported by any conclusive experimental evidence." The authors favoring this theory, which considers only one of the family of parastichies, must call upon Snow's theory in order to explain some observations. Loiseau (1969) discourses thoroughly upon Plantefol's theory. In 1974, Schwendener's contact pressure theory (1878) was strongly revitalized by Adler (1974, 1975, 1977a) who showed the limitations of Snow's and Richards' approaches.

At present there is no complete explanation of the phenomenon of phyllotaxis on the basis of contact pressures or chemical, nutritional, or morphogenetic fields. This may come from the fact that phyllotaxis is not a purely mechanistic or reductive problem. Chemical and mechanical factors are undoubtedly involved in the distribution of the primordia, but the mechanistic explanations finally demonstrate the need for a higher level of universality to deal with the problem. Several works on phyllotaxis, generally ignored by the mathematical approach, concerning ecology, phylogeny, and vascularization, must be taken into consideration. Phyllotaxis is an epigenetic, holistic, systemic phenomenon; it operates beyond chemistry and physics. The problem has three aspects: descriptive, mechanistic, and functional. The goals are to find an adequate description of the phenomenon, to determine parameters characterizing the systems, to create mechanisms by which we can generate the patterns, to provide likely models by which it becomes possible to control the phenomenon, and to relate the systems to their functions in their natural environments. The functional aspect has been almost completely neglected by the modelers. This aspect forces us to search for the phyletic and ecological origins of phyllotaxis. Corner (1981) asserts that the brown algae are the only plants that can unveil the mystery of the origins of phyllotaxis. General systems theory, where the notions of entropy and hierarchy play a major role, constitutes a synthetic approach to the problem. The theories presented in Chapter 5 are compared here, their successes are enhanced, and a few research orientations are pointed out.

6.2 DIFFUSION THEORY

The limits of the models of diffusion were perceived in Chapter 4, where we saw the lack of concentration gradient or of steady-state equilibrium in the various formulas used, in spite of the rapid transportation of substances within the apex. These models are based on the works of Richards, which contain, implicitly, assumptions describing the behavior of the inhibitor. They can be stated as follows (Adler, 1974, p. 29).

Assumption R1.   The level at which the new primordium arises is mainly determined by the inhibitor secreted by the stem apex. The concentration of this inhibitor decreases as the distance from the apical tip increases. The primordium arises at that level where the concentration has attenuated below a certain critical threshold.

Assumption R2.   At that level, the primordium arises where the sum of the concentrations of the inhibitor secreted by the existing leaf primordia is a minimum. The concentration of the inhibitor from a single leaf primordium is a decreasing function of its distance from the center of the primordium.

The nature of the inhibitor stays indeterminate. Of course the identification of an active inhibitor would be a step in the verification of the theory. According to Steeves and Sussex (1972) the field would finally result not from the creation of inhibiting substances, but from the withdrawal of essential nutritive elements. However, auxin (Indoleacetic acid) possesses many properties of the inhibitor but is unlikely to be the inhibitor sought. The diffusion mechanism is only a postulate; its existence has not been proved. The microsurgical experiments conceived to block the way of eventual diffusing substances are difficult to interpret, and some observations (see McCully and Dale, 1961), or experiments (see Snow and Snow, 1952), go against the diffusion theory. But this theory can generate the observed spirality from an initial distribution of the primordia, and the equalization of the consecutive divergences between these primordia.

Thornley's model (Section 5.4.1) does not always give a constant divergence angle or a series of divergence angles that settle rapidly around a precise value. The parameter λ of his model corresponds to nothing observable; it cannot be interpreted morphologically or physiologically. Thornley supposes that small perturbations arise within the system, so that the new primordium springs up at a minimum other than the absolute minimum of the concentration, which is, moreover, a variable threshold. He needs more than a dozen perturbations of the parameters to produce an angle that is approximately equal to ϕ^{-2}. The postulate of spatial competence leads Thornley to conclude that the origin of phyllotaxis must be sought in groups of cells, in tissues, rather than at the single cell level. For Church (1920, pp. 23 and 32), "The mechanism of Fibonacci phyllotaxis is ...completely independent of cell-segmentation,...the mechanism has no relation whatever to the more obvious cell-framework of the plant apex...." Veen and Lindenmayer's model (Section 5.4.2) precisely operates on that level. Their program can generate, from an initial configuration, phyllotactic patterns with low secondary numbers, such as 5/3, 5/2, 5/4, 4/3, and 7/1, and some approximations of the observed divergence angles. Young's model strikes against the same limits. These results are interesting, however, since these types of phyllotaxis are the most common, according to Fujita (1938).

But simulation is not the main concern in botanometry. The point is not just to ask the computer to go on reproducing a pattern outlined by an initial configuration. Two problems cannot be solved by the diffusion theory. The first one is to know what determines the spirality, that is, how the initial configuration of the primordia settles down. The second is to reproduce the phenomenon of rising phyllotaxis. Some claim that the initial arrangement is a consequence of Richards' theory (see Research Activity 1 in Section 5.6). Others argue that the increase of the apex diameter generates the 2/1 phyllotactic pattern, and they invoke the force imagined by van Iterson, recalling the plastic disks of Snow (1955), to explain the rising phyllotaxis.

6.3   CONTACT PRESSURE THEORY

The contact pressure hypothesis does not pretend to explain the initial placement of the primordia; it assumes a spiral arrangement where the consecutive divergences are equal to some value of d between $\frac{1}{3}$ and $\frac{1}{2}$. This assumption can be replaced (see Section 5.2) by the weaker hypothesis according to which the primordia lie on a fundamental spiral, and the successive divergences have values between $\frac{1}{3}$ and $\frac{1}{2}$. Adler also assumes that the rise r is a monotonic decreasing function of the time T. Contact pressures can then generate, as predicted by Schwendener, the phenomenon of rising phyllotaxis. The main result states that the convergence toward the divergence angle ϕ^{-2} is inevitable if contact pressure begins before leaf 6 appears, or r becomes smaller than $\sqrt{3}/38$; we then have the initial 2/1 or 3/2 pattern, and the rising of phyllotaxis along the main series follows if contact pressure is maintained. Thus Adler's model delivers conditions giving the initial 2/1 pattern.

This model gives a precise idea of the value of the divergence when the phyllotaxis of the system is expressed by high secondary numbers. For the systems presenting the phyllotaxes 2/1 and 3/2, the divergences given by the model are in relatively wide intervals, that is, [128.57°, 180°] and [128.57°, 142.11°], respectively (see Exercise 2.33). Van Iterson found that in seedlings of numerous species of plants possessing Fibonacci phyllotaxis, the first two primordia after the cotyledons arise at nearly 180° from one another, and that the divergence angles of the subsequent primordia do not approximate the normal angle 137.5° until several primordia have arisen; the process can be interpreted as a gradual approach to the above angle. Adler's model is consistent with this finding, since it predicts an oscillation of the divergence angle, with diminishing amplitude, around this value. On the other hand, the diffusion theory cannot account for the special role played by this angle. Moreover, Fujita (1939) took hundreds of measurements of angles on the apices of more than 30 species presenting the patterns 2/1 and 3/2. The observed divergences are in the above intervals, along approximately a normal curve centered at 137.5°. But the conditions under which Fujita's measurements were taken are imprecise; he considered above all the conspicuous (2, 1) and (3, 2) systems, leaving aside the limiting cases, near the triple points (1, 1, 2), (1, 2, 3), and (2, 3, 5), giving angles near the end points of the preceding two intervals.

Adler's model assumes, without trying to explain, the absence of a primordium in the distal zone of the apex (see Figure 5.2); it needs the bare apex postulate. For this reason it would not be, according to Richards (1951), a valid theory of phyllotaxis. Adler verified that if the distances between neighboring primordia are measured on the surface of a paraboloid, a cone, or a disk, then contact pressure does not give ϕ^{-2}. Moreover, the modifications idealized by Roberts (see Research Activity 3 in Section 5.6) for the contact pressure model are not completely general. Finally one can ask just how much of Adler's results can be obtained without mentioning continued fractions. I leave the question open to the reader's reflection. The answer is considered in Research Activity 1 of Section 5.6, and Jean's theorem of Section 1.3.2 can be straightforwardly generalized to visible pairs.

6.4   LIMITATIONS OF THE MECHANISTIC THEORIES

Richards argues that it follows from R1 and R2:

Assumption C1.   With the level of the new primordium determined mainly by the stem apex, its position at that level is determined above all by the two primordia that are its nearest neighbors.

Assumption C2.   The primordium is formed "tangentially somewhat nearer the older one" of these two primordia.

He then concludes the following from C1 and C2:

Assumption C3.   The successive divergences of the new leaf primordia converge toward the angle ϕ^{-2}.

Adler refuted the idea that R1 and R2 imply C3, by producing a counterexample showing that there exist functions satisfying the premises but requiring that the consecutive divergences converge to a limit that may be any number in the range [72°, 180°] (see Research Activity 7 in Section 5.6). This supports the original conjectures by Schoute (1913): an inhibitor field theory can explain why a leaf distribution tends to approximate a regular lattice (that is, to give successive divergence angles approximately equal along a genetic spiral), and the contact pressure theory can explain why the divergences tend to $(t + \phi^{-1})^{-1}$. This does not argue for the idea often put forward that "an inhibitor mechanism can be formally equivalent to contact pressure," in the sense that a short-range, long-lived inhibiting action would determine regions (around the leaves) acting very much like hard spheres.

Assumption C1 contains the ambiguous term "nearest neighbors." In the cylindrical lattice the meaning of this term is familiar, but this cannot be the one adopted by Richards, since we would know the neighbors of a primordium before it arises, which is equivalent to placing the primordium at an appropriate point. This term can have only the following interpretation:

Assumption S. The nearest neighbors are those that are nearest vertically, that is, that are on the first two levels below the new primordium, with regard to the tip of the apex.

Adler (1975, p. 443) remarked that C1, C2, and S transform Richards' theory into a special case of his model of Snow's theory. This underlines Snow and Snow's statement (1948) that the first available space theory is often confused with the repulsion (inhibitor) theory. The model in Section 5.3.3 shows that Snow's hypothesis can give a helical pattern of primordia and divergence angles tending to a limit d in the interval $(0, \frac{1}{2})$; that the angle ϕ^{-2} is just the result obtained by giving a particular value to the parameter a on which the divergence depends; and that for $\frac{1}{2} < a < \frac{2}{3}$, an assumption equivalent to C2, we obtain divergence angles producing the 2/1, 3/2, and 5/3 phyllotactic patterns. It follows that from C1, C2, and S it is not possible to deduce C3, but it is possible to obtain the above patterns.

The conclusion is finally reached that neither of the two rival theories of the 1940s (Snow's and Richards') has any advantage over the other. In some respects, Snow's or Richards' theory may be combined with Adler's, each being called on to supply what the other seems to lack. Recently Williams (1975) expressed his preference for a theory that would combine mechanical forces, such as pressure, and physicochemical processes, such as diffusion. This constitutes a most stimulating research orientation, where Turing's stationary wave theory could apply (Section 4.4.3). But according to Wardlaw's eighth principle of plant organization (1965b), the progressive organization of plants comes not only from the genes, but also from the environment. There are thus reasons to think that to solve the mystery, the context must be enlarged so as to comprehend the functional aspect of the problem. Sinnott (1960, p. 161) puts forward the idea that the usual theories are "too simple"; he looks for "more logical" solutions.

6.5   THE FUNCTIONAL PROBLEM

The phenomenon of phyllotaxis is obviously embedded in the natural environment. Discovering its relations to this environment constitutes the functional aspect of the problem of phyllotaxis. This has scarcely been worked on, and is now briefly presented here.

In the eighteenth century Bonnet argued that the perspiration of leaves demands that the air circulate freely among them, and that consequently they must cover each other as little as possible and adopt the spiral pattern. In 1871 Wright put forward the functional explanation that the need for light induces the minimal superposition of the leaves, and that this is provided by spirality and a divergence angle of ϕ^{-2} (see Research Activity 3 in Section 2.5). Wiesner (1907) claimed he had proved this assertion experimentally. What he proved, in fact, was that the lower leaves shaded by the upper leaves receive less light! According to many investigators (Church, 1904; Thompson, 1972; Richards and Schwabe, 1969) Wright's suggestion is not tenable; phyllotaxis is in fact a wide-ranging optimization problem. The photosynthetic argument opens the way, however, to the idea that Fibonacci phyllotaxis is the result of maximizing the global energy of plants, that is, of minimizing the entropy. This looks like a central unifying point of view for an epigenetic approach to the problem. Lehninger (1969) develops the point of view that a convenient study of biology should start with the principles of thermodynamics, which are "the most fundamental way to analyze all biological processes." When an inhibitor diffuses, there is an increase in entropy; when a primordium produces an inhibitor, there is a decrease. Plant-pollinator interaction is a factor of negative entropy. It is known, indeed, that the evolution of some plants is essentially controlled and guided by the selective activities and the sensorial capacities of pollinators searching for food. For many reasons (see Leppik, 1970), pollinators are particularly attracted by the system of numerical patterns of the capituli of the Compositae (such as the sunflower and the daisy), perfect examples of spiral phyllotaxis. Now the increase in the survival capacity due to plant-pollinator interactions is one of the main factors that determine the stability of the systems (see King, Gallaher, and Levin, 1975). By means of many experimental treatments it has been shown that the phyllotaxis of a plant can be greatly modified: "These results demonstrate conclusively that the phyllotaxis of an apex is dependent on circumstances and, if interfered with, may readily change from one system to another" (Richards, 1948, p. 219). In phyllotaxis the most significant problem is not to say "how it works," by taking the regulations for granted, but to explain how these regulations arise.

For Church the origin of Fibonacci phyllotaxis is phylogenetic, thus putting in the background the questions of photosynthesis and growth conditions. According to him, one should look for the origin of phyllotaxis in the sea, at the beginning of evolution. Facets of the problem that look important on the ontogenetic level can take their real places on a higher level. For example, in brown algae there is no problem of superposition of leaves, mentioned earlier in relation to photosynthesis. These algae nevertheless present elegant examples of Fibonacci arrangements. For Church (1904) the mechanism of phyllotaxis in higher plants is but the amplification of phytobenthic factors and it is, according to him, from this point of view that one should try to understand the arrangements of leaves on terrestrial plants. Whorled patterns would be secondary and phylogenetically related to the most primitive patterns, the spiral patterns. In that context it is not so important to try to explain superposed whorls, as in Thornley's model, which cannot in fact explain them.

The model in Section 5.5, a modest introduction to the functional aspect of the problem, uses hierarchies showing the rhythmic production of simple and double points, which looked so important to Church in phyllotaxis, producing an asymmetry responsible for the observed spirality. For Church (1920), "If there is no rhythm there is no pattern and conversely." Contact pressure determines a rhythm of growth, the rising of the phyllotaxis on the contact pressure path, and can be considered as one of the possible indications of the presence of the rhythm. Now works on vascular phyllotaxis, justifying the representation of the primordia in hierarchies, show (see Jean, 1982, p 750) on the sympodial level the rising phyllotaxis and the shrinking of the corresponding interval for the divergence, as in Adler's model (see Exercises 2.27 and 2.30). The main result of the interpretative model formulated in the hierarchical representation of the primordia is that normal Fibonacci phyllotaxis is inevitable if and only if the rhythm is induced before time $T_r = 6$ (compare with the main result in Adler's model; in Proposition 3 of Section 5.2.3, the leaf 0 is our leaf 1) or at T_r = 8 or 13.

Richter and Schranner (1978) suggest that all the types of phyllotaxis can be obtained from two main characteristics of the inhibiting interaction, namely, its lifetime and spatial range. More precisely, they assert that for a given spatial range, the increase in the duration w of the inhibiting activity would produce normal phyllotaxis (Section 2.2.2) for larger and larger values of t in Expression 2.1. In the hypothetico-deductive construction of Section 5.5, this is produced by increasing the time T_r at which the rhythm is induced. The identification of w and T_r brings a certain cohesion between the two points of view.

The systemic model, in spite of its limits, explains the predominance of small secondary numbers in Fibonacci phyllotaxis, and the predominance of this type of phyllotaxis in nature. It predicts the relative frequencies of appearance of the various patterns and shows which types can exist and which types cannot (according to Popper (1968), "Every 'good' scientific theory is a prohibition: it forbids certain things to happen"), thus opening the way to a verification of the model.

6.6   AN OVERVIEW OF THE MATHEMATICAL MODELS IN PHYLLOTAXIS

The section lists some of the contributors to botanometry and some of the biological theories on which the mathematical models are based.

Descriptive

1. Cylindrical (Chapters 1 and 2): Bravais (1837), van Iterson (1907), Coxeter (1972), Adler (1974), Jean (1981b), Erickson (1983).
2. Centric (Chapter 3): Church (1904), Richards (1951), Mathai and Davis (1974), Thornley (1975b), Thomas (1975), Jean (1979b).
3. Hierarchical (Section 5.5): Bolle (1939), Stewart (1964), Jean (1982).
4. Conical, Paraboloid (Section 3.4): van Iterson (1907), Adler (1977).

Explanatory

1. Mechanistic: (a) Chemical (computer simulation) (Section 5.4): Veen (1973), Thornley (1975a), Young (1978); (b) Physical (Sections 5.2, 5.3): Adler (1975, 1977).

2. Interpretative (functional) (Section 5.5): Jean (1980c).

Empirical

1. Sigmoid laws (Section 3.4): Richards (1969).
2. Allometric law (Section 3.3.5): Jean (1983e, 1983g).

Underlying Theories

1. Hofmeister's axiom (1868).
2. Schwendener's contact pressure theory (1878).
3. Church's phylogenetic theory (1904a, 1968).
4. Schoute's (1913) and Richards' (1951) diffusion theory.
5. Szabo's (1930) and Bolle's (1939) theory of bifurcating induction lines.
6. Snow and Snow's (1962) first available space theory.
7. Zimmermann's telome theory (Wilson, 1953, and Stewart, 1964).
8. Bare apex postulate (Adler, 1977).
9. Theory of the acropetal induction of the primordia by the foliar traces (Esau, 1943).
10. Principle of optimal design (Rosen, 1967).
11. Williams' mechanico-chemical field theory (1975).

6.7 CONCLUSION

From the time when the study of botany consisted of naming and classifying plants, Leonardo da Vinci, one of the first to be amazed by the regular patterns on plants, Nehemiah Grew, an eminent botanist who held that plants are an incentive for mathematical researches, and Goethe, the author of the famous theory of the metamorphosis of the leaves, are landmarks leading to Schimper and Braun's arithmetic conceptions and to Bravais' geometrical works. Botany today uses many sciences, and mathematics plays an important role on both phenomenological and causal levels. Botanometry has recently seen remarkable progress through mathematical modeling. It now offers a range of refinable models, suggesting directions for future researches.

But even the descriptive aspect of the problem of phyllotaxis must be revisited. Showing a way to attack a riddle is an even greater achievement than solving the riddle; this disturbs our laziness and complacency. What would be the state of development of quantum theory if all workers in the field had ever shown complete satisfaction with Bohr's model? Research in botanometry should aim at a better integration of the various aspects and treatments of the problem of phyllotaxis. Investigators interested in the field from a purely mathematical standpoint tend to pay insufficient attention to the relevant anatomical and physiological aspects, whereas those who are mainly interested in the biological aspects tend to minimize the geometric aspect. The growing apex is a geometric dynamic system possessing a biological organization. The empirical techniques of investigation must be refined and the theoretical models must become more embodied if the large gap between factual and theoretical phyllotaxis is to be filled.

From the point of view of allometry, phyllotaxis received very little, if any, mathematical consideration. Dormer (1965, p. 468) wrote that "Practically nothing is known concerning the correlations of discontinuous variables with each other, and very little more of their correlations with variables of other classes. Although the opportunities are restricted it is surprising that the subject should be so completely neglected. One should have expected at least that the extensive literature on meristic variation, especially in flowers, would have led to some examination of the correlations involved, but nothing significant seems to have emerged. It is difficult even to imagine the kind of relationships likely to arise, though presumably the Fibonacci numbers will occupy a specially prominent position." The results of Section 3.3.5 at least compel a revision of Dormer's sayings, except of course for the role of the main series. Those quite elegant and efficient allometric relations make Richards' phyllotaxis index, in fact a concept somewhat artificial, seem rather obsolete. It appears that future progress and the most valuable mathematical theories of phyllotaxis will come from a massive integration of the concepts expressing differential growth. Klein's path curves (Research Activity 11 in Section 5.6), which in certain cases give allometric equations, should be investigated. The various cylindrical lattices can be continuously transformed into one another, just as Bookstein's grids (1978) in morphometry allow one to reduce every change of form to gradients of differential or allometric growth. The study of relative growth rates in the apex will redirect the debate in the framework of Thompson's transformation theory and of the evolutionary implications of the subject of phyllotaxis.

Richards (1948, pp. 241–242) points out the fact that the theories are first formulated for specialized systems, then try to accommodate the aberrant cases: "The converse attitude would seem to provide a safer approach." Spiromonostichy, observed in *Costus*, is not a multijugate system (Research Activity 5 in Section 2.5), it "violates Hofmeister's rule" (Smith, 1941), but is in agreement with the theory of induction of leaves by the foliar traces (Research Activity 6 in Section 5.6). For Plantefol (1948, 1956), "It is the study of abnormalities that is fruitful for interpreting the phyllotaxis of a species and for eliminating invalid hypotheses." The mechanistic theories of phyllotaxis may be put in question when one sees, as in cases of natural transitions from decussation to distichy (see Guédès and Dupuy, 1983), a primordium initiated as close as possible to the preceding one, thus infringing the premises of physical and chemical inhibitory effects of the primordia. Disrupting normal patterns, with surgical experiments or with chemical treatments, may give clues on the nature of the control mechanism at work.

Because the mechanisms responsible for these geometric patterns are unknown, the best one can do to help solve the problem of phyllotaxis is to make the assumptions as precise as possible and to work out the mathematics for them, hoping that eventually observations will be forthcoming that can be used to distinguish between some of them, or to feed the dialectical process between them. The progress of research toward more universal levels of consideration will reveal more and more clearly the limits of the validity of the actual theories and models in botanometry, while incorporating their successes. Mathematical modeling in phyllotaxis and related phenomena deeply rooted in plant physiology and morphogenesis, as well as in ecology and phylogenesis, is still in its infancy. Phytomathematics is young indeed, but it will probably someday experience great success, such as that of mathematical physics.

> Ah, Sun-flower! Weary of time
> Who countest the steps of the Sun,
> Seeking after that sweet golden clime,
> Where the traveller's journey is done.
>
> (from William Blake's Songs of Experience)

APPENDIXES

It is a chastening thought that, after so many decades of work, we still know so little of the factors which determine phyllotaxis and related phenomena (Wardlaw, 1968a).

Appendix 1
SOLUTIONS OF THE EXERCISES OF CHAPTER 1

1.2 FUNDAMENTAL CONCEPTS

1.1 The phyllotaxis of a specimen is made of two consecutive terms of the sequence $\langle F(k) \rangle = \langle 1, 1, 2, 3, 5, 8, 13, \ldots \rangle$, defined by the recurrence relation

$$F(k + 1) = F(k) + F(k - 1), \qquad F(1) = F(2) = 1.$$

This is the main series, also known as the Fibonacci sequence. In a word, the phyllotaxis of the plant is one of the terms of the sequence

$$2/1, 3/2, 5/3, 8/5, 13/8, \ldots, F(k + 1)/F(k), \ldots,$$

called the main phyllotactic series.

1.3 Prove first that $\lim_{k \to \infty} F(k + 1)/F(k) = \phi$: from the relation $F(k + 1) = F(k) + F(k - 1)$, we get $F(k + 1)/F(k) = 1 + 1/[F(k)/F(k - 1)]$, and, at the limit, putting $\lim_{k \to \infty} F(k)/F(k - 1) = x$, we get $x^2 - x - 1 = 0$, an equation whose positive solution is $x = \phi$. This number, that is, the golden section, or Phidias' number, is a constant of nature, just like π the constant of Archimedes, i the constant of Tartaglia, e the constant of Euler, k the constant of Boltzmann, $\hbar$ the constant of Planck, or c the speed of light. Philosophizing on the presence of ϕ^{-2} and of the series $\langle F(k) \rangle$, botanists spoke of "a law of nature." For Coxeter it is "a fascinatingly prevalent tendency." We can at least speak of the "principle of phyllotaxis." "What a poem the analysis of phi" (ϕ), Valéry used to say.

1.4 The Table A.1 presents commonly encountered phyllotactic fractions, and the Introduction shows how to determine and to interpret them. The most usual fractions belong to the series

$$1/2, 1/3, 2/5, 3/8, 5/13, 8/21, \ldots, F(k)/F(k + 2), \ldots$$

known as Schimper and Braun's series, after the investigators who discovered this preference for the main series.

Table A.1 Phyllotactic fractions of common species (from Jean, 1978b)

Phyllotactic fractions	Species
1/2	Elm, linden, mulberry
1/3	Adler, beech, hazel, birch, yucca
2/5	Plum, oak, cherry, apple, apricot
3/8	Plantain, poplar, pear, sycamore
5/13	Willow, almond, white pine

1.6 For the positive root ω of the equation $x^2 = ax + 1$ or $x = a + 1/x$, $a \geq 1$, we obtain $\omega = [a; a, a, a, \ldots]$. Setting $a = 1$, $\omega = \phi$, and $a = 2$ gives $\omega = \sqrt{2}$. These remarkable formulas connect ϕ and $\sqrt{2}$ to the integers in a much more striking way than the decimal expansions of these numbers do, since the latter display no regularity in the succession of their digits.

1.7 Considering the laws of formation of the convergents (Expression 1.3), multiply the first relation by q_{k-1}, the second by p_{k-1}, and subtract the second from the first to obtain

$$q_k p_{k-1} - p_k q_{k-1} = -(q_{k-1} p_{k-2} - p_{k-1} q_{k-2}).$$

Since $q_1 p_0 - p_1 q_0 = -1$, the result follows.

1.8 Verify, in the case of ϕ, that the assumption is true for the first few terms of the series, and then prove the result using the laws of formation of the convergents. In the case of $1/\phi^2$, use the relation $1/\phi^2 = 1/(1 + \phi)$.

1.10 Using the answer to Exercise 1.6 with $a = 4$, we rapidly find $11/5, 20/9, 29/13, 47/21, 85/38, 123/55, \ldots$. Since $1/(e - 1) = [0; 1, 1, 2, 1, 1, 4, \ldots]$, $2/3$ and $11/19$ are intermediate convergents.

1.11 For the second part, $(3, 2)$ and $(19, 11)$. Compare Exercises 1.10 and 1.12.

1.12 The parametric equation of the segment is, for the real number c:

$$x = \frac{c}{a_n} q_n + \left(1 - \frac{c}{a_n}\right) q_{n-2} = c q_{n-1} + q_{n-2},$$

$$y = \frac{c}{a_n} p_n + \left(1 - \frac{c}{a_n}\right) p_{n-2} = c p_{n-1} + p_{n-2}.$$

Between the two vertices, $0 < c < a_n$, and the points of the lattice between these vertices correspond to the integral values of c. It follows, in the last case, that $(x, y) = (q_{n,c}, p_{n,c})$.

1.3   APPROXIMATION FORMULA FOR THE DIVERGENCE

1.13   With the orthostichy $0, 34, 68, 102, \ldots$ we have $c = 3$, $s = 2$, $cm + sn = 34$, $cp + sq = 13$, and $d \simeq 13/34$.

1.16   Refer to Figure 1.5. Point $C = 16$ can be reached from point $a = 0$ or from point $D = 1$: $mq = np + 1$. Point $Q = 25$ can be reached from point $D = 1$ or from point $A = 0$: $mq = np - 1$.

1.17   Suppose that $\overline{CD}$ meets $\overline{aA}$ in R, $\overline{QD}$ meets $\overline{aA}$ in S, and T is the foot of the perpendicular from D on $\overline{aA}$. By the properties of the similar triangles aCR and aBA we have

$$\frac{\overline{aC}}{nl} = \frac{\overline{CR}}{ml*} = \overline{aR}$$

where l and $l*$ are defined in the text. It follows that

$$\frac{q}{n} = \frac{pl* + \overline{DR}}{ml*} = d + \overline{TR} \qquad \text{(A.1)}$$

and $d \le q/n$. Consider now the similar triangles DSR and aBA. We have

$$\frac{\overline{DR}}{ml*} = \overline{SR} = \overline{ST} + \overline{TR},$$

so that $\overline{TR} \le \overline{DR}/ml*$. From Expression (A.1) we get $d \ge p/m$ (where $mq - np = 1$). Working with the triangles SQA and aBA, we obtain $q/n \le d \le p/m$ (where $mq - np = -1$).

1.18   By Exercise 1.17 we have

$$p/S_{t,k} \le d \le q/S_{t,k-1}, \text{ or } q/S_{t,k-1} \le d \le p/S_{t,k},$$

where $mq - np = \pm 1$, that is, $S_{t,k}q - S_{t,k-1}p = \pm 1$. The only solution is $(q, p) = (F(k - 1), F(k))$, since

$$F(k - 1)^2 - F(k)F(k - 2) = \pm 1,$$

according to whether k is odd or even. By the principle of closed nested intervals, the intersection of the intervals under consideration contains only one point, that is, $d = \lim_{k \to \infty} F(k)/S_{t, k} = (t + \phi^{-1})^{-1}$.

1.4 A FIRST EXPLANATION IN PHYLLOTAXIS

1.20 The abscissa of $F(k)$ in the cylindrical lattice is

$$\phi F(k) - F(k + 1) - [\phi F(k) - F(k + 1)].$$

In the expression $\phi F(k) - F(k + 1)$, replace $F(k)$ and $F(k + 1)$ by their values given by *BINET'S FORMULA*:

$$F(k) = \frac{\phi^k - (-\phi)^{-k}}{\sqrt{5}},$$

to obtain $\phi F(k) - F(k + 1) = (-1)^{k+1}(\phi^2 + 1)/\sqrt{5}\,\phi^{k+1}$. Since $\phi^2 + 1 = \sqrt{5}\,\phi$, it follows that $[\phi F(k) - F(k + 1)]$ is equal to 0 or to -1. Thus the coordinates of the neighbor $F(k)$ of the Y-axis in the cylindrical lattice are as expected. Now consider, in the (X, Y)-plane, the square of the Euclidean distance from the origin to $F(k)$, and to $F(k + 1)$, two neighbors of the Y-axis: $1/\phi^{2k} + F(k)^2 r^2$ and $1/\phi^{2k+2} + F(k + 1)^2 r^2$, respectively. Notice that the second expression is smaller than the first for $r^2 < 1/\phi^{2k+1}[F(k + 1)^2 - F(k)^2]$. Finally the absolute value of the abscissa of $F(k)$ is greater than the absolute value of the abscissa of $F(k + 1)$.

1.21 We must have $d(0, 3) < d(0, 2)$, and $d(0, 5) < d(0, 8)$, where $d(0, x)$ is the Euclidean distance from the origin to point $x = (x_1, x_2)$, that is, $(x_1^2 + x_2^2)^{1/2}$. The coordinates of the points 2, 3, 5, and 8, as given in the theorem of Section 1.4.1, correspond to $k = 3$, 4, 5, and 6, respectively. It follows that

$$\frac{1}{\sqrt{39}}\phi^{11/2} < r < \frac{1}{\sqrt{5}}\phi^{7/2}.$$

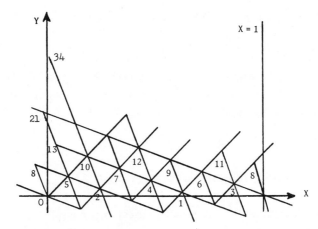

Figure A.1 Point 8 has become a neighbor of the origin and the phyllotaxis has risen from 5/3 to 8/5 (from Jean, 1978b).

The phyllotaxis changes when $d(0, 8) = d(0, 3)$, that is, for $r = 5^{1/4}/\phi^5$ $\sqrt{55}$. Figure A.1 was drawn with this value of r. The phyllotaxis passes from $F(k)/F(k - 1)$ to $F(k + 1)/F(k)$ for $r = 5^{1/4}/\phi^k$ $(F(k + 1)^2 - F(k - 1)^2)^{1/2} = 5^{1/4}/\phi^k F(2k)^{1/2}$.

1.22 Suppose that $d = k_1/k_2$ is a rational number. Leaf k_2 is then on the Y-axis since its abscissa is given by $k_2 d - [k_2 d] = 0$.

1.23 For $n \geq 2$, $q_{n, a_n - 1} = (a_n - 1)q_{n-1} + q_{n-2} = q_n - q_{n-1}$, and $q_{n+1, 1} = q_n + q_{n-1}$. See the definition of the intermediate convergents and the laws of formation of the convergents in Section 1.2.3 and Exercise 1.12. It follows that $q_{n-1} < q_{n,1}$ and $q_{n, a_n - 1} < q_n$. Since $q_{n, i} < q_{n, i+1}$ the result becomes obvious.

1.24 The theorem in Section 1.4.1 gives the coordinates of the points $F(k)$ and $F(k + 1)$. It is well-known that two points (x_1, x_2) and (y_1, y_2) make a right angle with $(0, 0)$ when $x_1 y_1 + x_2 y_2 = 0$. It follows that $-\phi^{-2k-1} + F(k)F(k + 1)r^2 = 0$.

1.25 Draw the visible opposed parastichy triangle belonging to the pair 7/4 and use Bravais' approximation formula to get

$$cm + sn = 7c + 4s = S(k),$$

where $S(k)$ is the kth term of Schoute's accessory series, and $(c, s) = (F(k - 3), F(k - 4))$. Since $p = 2$ and $q = 1$ $(7q - 4p = \pm 1$:

Exercise 1.16), we have $cp + sq = F(k - 1)$. The result follows from the relations $d = \lim_{k \to \infty} F(k - 1)/S(k)$, and $S(k) = F(k + 1) + F(k - 1)$. Following the genetic spiral we have to come back five times to the left to meet 18 points: the phyllotactic fraction has the value $F(5)/S(6) = 5/18$. More generally, with the point $S(k)$ the fraction is $F(k - 1)/S(k)$.

1.26 By Binet's formula (Exercise 1.20),

$$\frac{S_{t,k}\phi}{\phi t + 1} = \frac{\phi^k}{\sqrt{5}} + \frac{(-1)^k(\phi - t)}{\phi^{k-1}(\phi t + 1)\sqrt{5}},$$

and $\phi^k = \sqrt{5} F(k) + (-1)^k/\phi^k$. It follows that the above expression has the value

$$F(k) + \frac{(-1)^k(\phi + 2)}{\phi^k(\phi t + 1)\sqrt{5}}.$$

Since $\phi + 2 = \sqrt{5}\,\phi$, we have the expected result.

1.27 The distance of $(S_{t,k}, F(k))$ to $y = (t + \phi^{-1})^{-1} x$ is obtained by putting the line in its normal form. This distance is $K|\phi F(k - 1) - F(k)|/(\phi t + 1)$, where K is a constant, and Binet's formula gives that the absolute value is ϕ^{-k+1}. Exercise 1.26 gives the other distance.

Appendix 2
SOLUTIONS OF THE EXERCISES OF CHAPTER 2

2.2 FUNDAMENTAL CONCEPTS

2.1 Given $m > n$, we have $d(0, m) \leq d(0, n)$, that is, $x_n^2 + n^2 r^2 \geq x_m^2 + m^2 r^2$ or $(m^2 - n^2) r^2 \leq x_n^2 - x_m^2$ only if $x_n^2 - x_m^2 \geq 0$ or $D_m \leq D_n$.

2.2 We have $4/5 < 2d < 1$, $D_2 = 1 - 2d < D_1$, and $n_2 = 2$; also $6/5 < 3d < 3/2$, and $D_3 = 3d - 1 > D_2$, so that $n_3 \neq 3$. Moreover, we know that in the closed interval $[1/3, 1/2]$, $n_4 \neq 4$. Finally, $2 < 5d < 5/2$, and $D_5 = 5d - 2$ will be smaller than D_2 for $d < 3/7$.

2.3 In this interval the points of close return are 1, 2, 3, 5, and 8, and the next close return candidate is 13. By Exercise 2.1, these leaves can become neighbors of the origin since $D_8 < D_5 < D_3 < D_2 < D_1$, and they correspond to the sequence of normal phyllotaxis with $t = 2$.

2.4 From Exercise 2.2, the points 1, 2, and 5 are points of close return. It is easily seen that $D_6 > D_5$ and $D_7 < D_5$, so that 7 is a point of close return. Also $D_{12} < D_7$ and 19 is the next close return candidate. By Exercise 2.1 the leaves 1, 2, 5, 7, 12, and 19 can become neighbors of the origin so that the system can display an anomalous type of rising phyllotaxis with $t = 2$.

2.5 Each intermediate convergent in the exercise is the mediant between the fraction preceding it and p_{n-1}/q_{n-1}; it is thus between these two fractions. It follows that for n fixed, the members of the given sequence are all smaller or greater than ω, a sequence decreasing in the first case, and increasing in the second. In order to determine a_n and p_n/q_n, the said sequence is built up to the point that the value of one of its members is on the other side of ω with regard to p_{n-2}/q_{n-2}; the last member on the same side of ω is p_n/q_n, which yields the value of a_n.

2.6 $\omega = (1, 2, 1, 2, 1, \ldots)$ determines the intervals $[0/1, 1/0]$, $[1/1, 1/0]$; $[1/1, 2/1]$, $[1/1, 3/2]$; $[4/3, 3/2]$; $[4/3, 7/5]$, $[4/3, 11/8]$, $\ldots$, and from the proof of the theorem in Section 2.2.2 we have $p_0/q_0 = 1/1$, $p_1/q_1 = 3/2$, $p_2/q_2 = 4/3$, $p_3/q_3 = 11/8, \ldots$, and $p_{1,1}/q_{1,1} = 2/1$, $p_{3,1}/q_{3,1} = 7/5, \ldots$.

2.7 The intervals are $[0/1, 1/0]$, $[1/1, 1/0]$; $[1/1, 2/1]$, $[1/1, 3/2]$; $[4/3, 3/2]$, $[7/5, 3/2]$, $[10/7, 3/2]$; etc, so that $\omega = 10/7$.

2.8 Since the first intervals of ω are $[0/1, 1/0]$, $[1/1, 1/0]$, $[2/1, 1/0], \ldots, [a_0/1, 1/0]$, $[a_0/1, (a_0 + 1)/1]$, we have $a_0 \leq \omega \leq a_0 + 1$.

2.3 DIVERGENCE ANGLES VERSUS VISIBLE OPPOSED PARASTICHY PAIRS

2.9 If $m > n$, then (m, n) is an extension of $(n + r, n)$, where r is the remainder of the division of m by n. If $r = 1$ the result follows. If $r \neq 1$, (m, n) is an extension of (r, n) where $n > r$, and the process starts again with the pair (r, n). There will be a step where the rest will be 1. Then the visible pair will be $(t, t + 1)$ if the genetic spiral winds up from left to right, or $(t + 1, t)$ if the genetic spiral goes from right to left. In both cases $td \leq 1 \leq (t + 1)d$.

2.10 Notice that the closed interval $[1/(t + 1), 1/t]$ and $\Delta(t, t + 1)$ have the properties stated in the fundamental theorem on the visibility of the extensions. The result follows. One more step delivers the following results:

$(3t + 2, t + 1)$ is visible if and only if

$$1/(t + 1) \leq d \leq 3/(3t + 2),$$

$(2t + 1, 3t + 2)$ is visible if and only if

$$3/(3t + 2) \leq d \leq 2/(2t + 1),$$

$(3t + 1, 2t + 1)$ is visible if and only if

$$2/(2t + 1) \leq d \leq 3/(3t + 1),$$

$(t, 3t + 1)$ is visible if and only if

$$3/(3t + 1) \leq d \leq 1/t.$$

2.11 $(2, 3)$ is visible if and only if $1/3 \leq d \leq 1/2$,
$(5, 3)$ is visible if and only if $1/3 \leq d \leq 2/5$,

(5, 8) is visible if and only if $3/8 \leq d \leq 2/5$,

(13, 8) is visible if and only if $3/8 \leq d \leq 5/13$.

For the other part of the exercise start with a right extension of $(2, 3)$.

2.14 In the case of the sequence in Expression 2.2, if $R_{t,k} = 2F(k-1) + F(k) + F(k-2)$, then n and m are two consecutive terms $R_{t,k}$ and $R_{t,k+1}$ of the sequence if and only if d is between the bounds $S_{t,k-1}/R_{t,k}$ and $S_{t,k}/R_{t,k+1}$; at the limit $d = [2 + (t + \phi^{-1})^{-1}]^{-1}$. For the other sequence, if $R_{t,k} = F(k+1)t + F(k-1)$, n and m are two consecutive terms $R_{t,k}$ and $R_{t,k+1}$ if and only if d is between the bounds $F(k+1)/R_{t,k}$ and $F(k+2)/R_{t,k+1}$; at the limit $d = (t + \phi^{-2})^{-1}$. In the latter case, start with the result $1/(t+1) \leq d \leq 1/t$ if and only if $(t, t+1)$ is visible; write the sequence of alternate extensions of $(t, t+1)$, beginning with a right extension.

2.15 The continued fraction for $d = 5/12$ is $[0; 2, 2, 2]$. Then we have $d = (0, 2, 2, 2, \infty)$ or $d = (0, 2, 2, 1, 1, \infty)$. The intervals in the first mediant nest are $[0/1, 1/0]$, $[0/1, 1/1]$, $[0/1, 1/2]$; $[1/3, 1/2]$, $[2/5, 1/2]$; $[2/5, 3/7]$, $[2/5, 5/12]$; $[7/17, 5/12]$,.... In the other case the mediant nest is the same up to $[2/5, 3/7]$; followed by $[5/12, 3/7]$; $[5/12, 8/19]$,.... In both cases $(2, 3)$ is a visible pair since $1/3 \leq d \leq 1/2$. The (visible) contractions are $(2, 1)$ and $(1, 1)$. The (visible) extensions, in the first case, are $(2, 5)$, $(7, 5)$, $(12, 5)$, $(12, 17)$, $(12, 29)$, followed by consecutive right extensions. In the other case the extensions are $(2, 5)$, $(7, 5)$, $(7, 12)$, $(19, 12)$, $(31, 12)$, followed by consecutive left extensions.

2.16 The continued fraction for $d = \sqrt{3}/6$ is $d = [0; 3, 2, 6, 2, 6, 2, \ldots]$. Then $d = (0, 3, 2, 6, 2, 6, 2, \ldots)$ and the intervals are $[0/1, 1/0]$; $[0/1, 1/1]$, $[0/1, 1/2]$, $[0/1, 1/3]$; $[1/4, 1/3]$, $[2/7, 1/3]$; $[2/7, 3/10]$, $[2/7, 5/17]$, $[2/7, 7/24]$, $[2/7, 9/31]$, $[2/7, 11/38]$, $[2/7, 13/45]$; $[15/52, 13/45]$,.... The pair $(3, 4)$ is visible since $1/4 \leq d \leq 1/3$. The contractions $(3, 1)$, $(2, 1)$, $(1, 1)$ are visible. The extensions $(3, 7)$; $(10, 7)$, $(17, 7)$, $(24, 7)$, $(31, 7)$, $(38, 7)$, $(45, 7)$; $(45, 52)$, $(45, 97)$,... are visible.

2.17 The series of extensions from $(0, 1)$ to $(17, 39)$ is $(0, 1)$; $(1, 1)$, $(2, 1)$; $(2, 3)$, $(2, 5)$; $(7, 5)$, $(12, 5)$, $(17, 5)$; $(17, 22)$, $(17, 39)$. This means that $a_0 = 0$, $a_1 = 2$, $a_2 = 2$, $a_3 = 3$, $a_4 \geq 2$, since the mediant nest is $[0/1, 1/0]$; $[0/1, 1/1]$, $[0/1, 1/2]$; $[1/3, 1/2]$, $[2/5, 1/2]$; $[2/5, 3/7]$, $[2/5, 5/12]$, $[2/5, 7/17]$; $[9/22, 7/17]$, $[16/39, 7/17]$, and $16/39 \leq d \leq 7/17$.

The principle convergents are $p_0/q_0 = 0/1$, $p_1/q_1 = 1/2$, $p_2/q_2 =$ 2/5, $p_3/q_3 = 7/17$. If $a_4 = 2$, $p_4/q_4 = 16/39$; if $a_4 = 3$, $p_4/q_4 =$ 23/56; if $a_4 = 2$ and $a_5 = 1$, $p_5/q_5 = 23/56$, etc.

2.18 $d = (0, 4, 2, 1, \ldots)$ and the mediant nest of intervals is $[0/1, 1/0]$; $[0/1, 1/1]$, $[0/1, 1/2]$, $[0/1, 1/3]$, $[0/1, 1/4]$; $[1/5, 1/4]$, $[2/9, 1/4]$; $[2/9, 3/13]$; .... The visible pairs are $(0, 1)$; $(1, 1)$, $(2, 1)$, $(3, 1)$, $(4, 1)$; $(4, 5)$, $(4, 9)$; $(13, 9)$; .... The convergents are $p_0/q_0 = 0/1$, $p_{1,1}/q_{1,1}$ $= 1/1$, $p_{1,2}/q_{1,2} = 1/2$, $p_{1,3}/q_{1,3} = 1/3$, $p_1/q_1 = 1/4$, $p_{2,1}/q_{2,1} =$ $1/5$, $p_2/q_2 = 2/9$, $p_3/q_3 = 3/13\ldots$.

2.19 The important word in the statement is "consecutive." Starting with $(0, 1)$ and the values of $a_1, a_2, a_3, \ldots$ obtained from the consecutive contractions of (m, n), the result of making a_1 left extensions is $(a_1, 1) = (q_1, q_0)$. Then making a_2 right extensions gives $(a_1, a_1 a_0 + 1) = (q_1, q_2)$. With a_3 left extensions we obtain $(a_3 q_2 + q_1, q_2) =$ (q_3, q_2). Continuing in this way it is obvious that the visible pairs that precede a change of direction are determined by a law of formation of the convergents, namely, $q_n = a_n q_{n-1} + q_{n-2}$.

2.4 PHYLLOTAXIS OF A SYSTEM

2.20 (a) Use induction.

(b) The two solutions of $np - mq = \pm 1$ for $n = F(k)$ and $m = F(k + 1)$ are (1) for $p = F(k)$ and $q = F(k - 1)$, $np - mq = (-1)^{k+1}$, and (2) for $p = F(k - 1)$ and $q = F(k - 2)$, $np - mq = (-1)^k$. If $x_m > 0$, we have in the first case, $D_m = F(k + 1)d - F(k)$ and $D_n = F(k - 1) - F(k)d$, and in the second case, $D_m = F(k + 1)d - F(k - 1)$ and $D_n = F(k - 2) - F(k)d$. Calculating the value of $(m - n)D_n + (2n - m)D_m$ we have in the first case, $(-1)^k(1 - d)$, that is, $1 - d$ if k is even, and d if k is odd, and in the second case, $(-1)^{k-1}d$, that is, $1 - d$ if k is even and d if k is odd. If $x_m < 0$, the results are as follows: in the first case we have $(-1)^{k-1}(1 - d)$; that is, d if k is even and $1 - d$ if k is odd. In the second case the expression has the value $(-1)^{k-2}d$; that is, d if k is even, and $1 - d$ if k is odd.

(c) The expression $(m - n)D_n + (2n - m)D_m$ is a monotonic function of d, increasing or decreasing according to the parity of k. The bounds are obtained by putting $D_m = 0$ and $D_m = D_n$. In the first case the bounds are n/m and $m/(m + n)$, and in the second case they are $(m - n)/m$ and $n/(m + n)$; both cases are for $x_m > 0$ and k odd, or for $x_m < 0$ and k even. Notice that in the first case, $1 - d$ is between $(m - n)/m$ and $n/(m + n)$, and in the second case, $1 - d$ is between n/m and $m/(m + n)$; both cases are for $x_m > 0$ and k even, or for $x_m < 0$ and k odd. If the divergence is smaller than $\frac{1}{2}$, then it must be within the bounds given in the exercise, since n/m and $m/(m + n)$ are greater than $\frac{1}{2}$.

(d)   We have $D_m = C|q_i d - p_i|$, and $D_n = C|q_{i-1} d - p_{i-1}|$. Performing the operations as in Proposition 2 of Section 2.4.1 yields the first result. Reworking part (b) above yields the other result.

(e)   The bounds obtained for the visible pairs are given by a theorem in Section 2.3.1: $m = F(k + 1) = S_{t, k-1}$ for $t = 2$, and $n = F(k) = S_{t, k-2}$ for $t = 2$. The bounds are thus $F(k - 1)/F(k + 1)$, and $F(k - 2)/F(k)$, that is, $(m - n)/m$ and $(2n - m)/n$. But $n/(m + n)$ given in part (c) is the mediant between the two bounds just obtained.

2.21   Corollary of the fundamental theorem on points of close return.

2.22   In particular we have

$$\frac{1}{3} < \frac{D_{2i+1}}{D_{2i-1}} < \frac{1}{2}, \qquad \text{for} \quad i = 1, 2, 3, \ldots, j.$$

Make an induction on j.

2.24   The opposed pairs that can be conspicuous are the pairs (n_{k-1}, n_k), $k = 2, 3, 4, \ldots$, where $n_k = S_{t, k-1}$. When r decreases, the pair (n_{k-2}, n_{k-1}) ceases to be conspicuous, leaving the place to the new conspicuous pair (n_k, n_{k-1}), as soon as $d(0, n_{k-2}) = d(0, n_k)$. This happens when

$$r = r_k = \frac{d\phi^{-k+1}(\phi^4 - 1)^{1/2}}{(S_{t, k-1}^2 - S_{t, k-3}^2)^{1/2}}.$$

Again, the pair (n_k, n_{k-1}) ceases to be conspicuous when $d(0, n_{k-1})$ $= d(0, n_{k+1})$. This happens when $r = r_{k+1} = d\phi^{-k}(\phi^4 - 1)^{1/2}/$ $(S_{t,k}^2 - S_{t,k-2}^2)^{1/2}$. So the pair (n_k, n_{k-1}) or the pair (n_{k-1}, n_k) is conspicuous, that is, the phyllotaxis of the system is n_k/n_{k-1}, when r belongs to the interval determined by r_k and r_{k+1}. Notice that for $t = 2$, $r_k = 5^{1/4}/\phi^k F(2k)^{1/2}$, as given by Exercise 1.21.

2.25 Use Propositions 2 and 3 in Section 2.4.1, and Exercise 2.20d. Notice that $y_m \simeq \phi y_n$ and $y_{m+n} \simeq \phi^2 y_n$.

2.26 Considering $\Delta(m, n)$, by the law of cosines $C^2 = n^2 d^2(0, m) + m^2 d^2(0, n) - 2mn d(0, m)d(0, n)\cos\theta$, where $\theta = 120°$.

2.27 The maximum of D_m/D_n occurs at the transition from m/n to $(m + n)/m$ phyllotaxis, where $d(0, m) = d(0, n) = d(0, m + n)$.

2.28 The algorithm at the end of Section 2.2.1 gives the answer to the cases (a), (b), (c), and (e):
(a) 1 and 2 are points of close return: arc $(2, 1)$.
(b) 1, 2, and 3 are points of close return: arc $(2, 1)$ and arc $(3, 2)$.
(c) arc $(2, 1)$: consider cases (a) and (b).
(e) 1, 2, 3, and 5 are points of close return: arc $(2, 1)$, arc $(3, 2)$, and arc $(5, 3)$ are relevant.
The answer to case (d) is given in Exercise 2.2: 1, 2, and 5 are points of close return, so that we can speak about the arc $(2, 1)$ and the arc $(5, 2)$. For the case (f), notice that $(5, 6)$ is a visible opposed parastichy pair (Exercise 2.9). It follows that the pairs $(0, 1)$, $(1, 1)$, $(2, 1)$, $(3, 1)$, $(4, 1)$, $(5, 1)$, and $(5, 6)$ are visible. Then 5 and 1 are denominators of principal convergents, by the Proposition of Section 2.3.2. They are denominators of consecutive principal convergents by the Theorem in the same section (one can also apply Exercise 2.19). The beginning of Section 2.4.1 tells us that 1 and 5 are consecutive points of close return. Thus arc $(5, 1)$ is the arc known to exist in case (f). The other cases in this problem can be dealt with the same algorithm. The theorem in Section 2.4.1 can also be used.

2.29 The value of the radius of the circle on which the arc rests is easily obtained from the two integers defining the arc and from Proposition 3 of Section 2.4.2. The well-known relation $p_i q_{i-1} - p_{i-1} q_i = (-1)^{i+1}$, $i > 0$, determines the values of p_i and p_{i-1} (rejecting the cases where $d > \frac{1}{2}$), and thus the center of the circle. Table A.2

summarizes the results:

Table A.2   The analysis of the first few arcs

Arc	Center	Radius	Permitted interval for d
(2, 1)	(2/3, 0)	1/3	[1/3, 1/2]
(3, 2)	(1/5, 0)	1/5	[1/3, 2/5]
(5, 2)	(8/21, 0)	1/21	[2/5, 3/7]
(5, 3)	(7/16, 0)	1/16	[3/8, 2/5]

2.30 (a) By Exercise 2.28, the points of close return are 1, 2, 3, and 5. The maximal value is given by the intersection of the arc (5, 3) with the arc (3, 2) (consult Figure 2.9). We easily obtain $r = \sqrt{3}/38$, and at this point we have $d(0, 5) = d(0, 3) = d(0, 2) < d(0, 1)$, and $d = 15/38$.

(b) In that case, 1, 2, and 5 are the points of close return. The value of r is on the arc (5, 2) with $d = 2/5$, that is, $r = 1/5\sqrt{21}$, where $d(0, 5) = d(0, 2) < d(0, 1)$.

(c) We must choose between the value $\sqrt{3}/38$ and the value $1/5\sqrt{21}$: $\sqrt{3}/38$. For leaf 2 we put $d(0, 2) = d(0, 1)$ with $d = 1/2$, to obtain $r = \sqrt{3}/6$.

2.31 The formula is

$$2\pi d = \frac{\epsilon\left(d^2(0, m)(m^2 - n^2) + 4\pi^2\right)}{4\pi mn} + \frac{\Delta_n}{2\pi n}.$$

Working with $\delta = 2\pi d$ we have $d^2(0, m) = (m\delta - 2\pi p)^2 + m^2 r^2$, $d^2(0, n) = (n\delta - 2\pi q)^2 + n^2 r^2$, with $np - mq = \pm 1$.

2.32 This amounts to calculating $d^2(0, m)$ in Proposition 2 of Section 2.4.2, using Exercise 2.20 and considering the fact that $d < \frac{1}{2}$ [$q = F(k - 2)$, $p = F(k - 1)$].

2.34 From Proposition 4 of Section 2.4.1, the pair (21, 58) is visible, and the divergence is in the interval $11/58 < d < 4/21$ (see the algorithm at the beginning of Section 2.3.2). Since 21 and 58 are denominators of principal convergents, by the proposition of Section 2.3.2, the pair

(21, 58) immediately precedes a change in the direction of the extensions, so that the next extension is (79, 58). Leaf x, on the arc (21, 58), such that $d(0, x) = d(0, 21) = d(0, 58)$, is a point of close return on the same side of the r-axis as leaf 21. Those leaves are linked by the relation $x = 58c + 21$ for consecutive denominators of principal convergents. Since there is no point of the lattice on the segment joining leaf 21 and leaf x (otherwise such a point would have been at the same distance from the origin, as leaf 21 and leaf 58, before leaf x), it follows that $c = 1$ and $x = 79$ (notice that the three leaves are related by the recurrence law defining the main series). The pair (79, 58) is thus visible and the corresponding interval of values for d is the one expected. More generally we have $x = m + n$.

REFERENCES

The phenomenon of plant growth constitutes so vast and diffuse a field of exploration that it would be presumptuous to pretend to give an adequate view of it in a book of a convenient size.

Abdulnur, S. F., Laki K., (1983). A Two-Dimensional Representation of Relative Orientations of α-Helix Residues, *J. Theor. Biol.*, **104**, 599–603.

Adler, I. (1974). A Model of Contact Pressure in Phyllotaxis, *J. Theor. Biol.*, **45**, 1–79.

Adler, I. (1975). A Model of Space Filling in Phyllotaxis, *J. Theor. Biol.*, **53**, 435–444.

Adler, I. (1977a). The Consequences of Contact Pressure in Phyllotaxis, *J. Theor. Biol.*, **65**, 29–77.

Adler, I. (1977b). An Application of the Contact Pressure Model of Phyllotaxis to the Close Packing of Spheres around a Cylinder in Biological Fine Structure, *J. Theor. Biol.*, **67**, 447–458.

Adler, I. (1978). A Simple Continued Fraction Represents a Mediant Nest of Intervals, *Fib. Q.*, **16**, 527–529.

Allsopp, A. (1964). Shoot Morphogenesis, *Ann. Rev. Plant Physiol.*, **15**, 225–254.

Allsopp, A. (1965). Heteroblastic Development in Cormophytes, in *Encyclopedia of Plant Physiology*, Vol. 1, W. Ruhland, Ed., Springer-Verlag, New York, pp. 1172–1221.

Ayala, F. J. (1974). The Autonomy of Biology as a Natural Science, in *Biology, History and Natural Philosophy*, A. D. Breck and W. Yourgrau, Eds., Plenum-Rosetta, New York, pp. 1–16.

Bailey, N. T. J. (1964). *The Elements of Stochastic Processes*, Wiley, New York.

Ball, E. (1944). The Effects of Synthetic Growth Substances on the Shoot Apex of *Tropaeolum majus* L., *Am. J. Bot.*, **31**, 316–327.

Ball, W. W. R. (1967). *Mathematical Recreations and Essays*, Macmillan, London.

Bamford, C. H., Brown, L., Elliott, A., Hanby, W. E., and Trotter, I. F. (1954). Alpha—and Beta—Forms of Poly-*l*-Alanine, *Nature*, **173**, 27.

Bard, J., and Lauder, I. (1974). How Well does Turing's Theory of Morphogenesis Work?, *J. Theor. Biol.*, **45**, 501–531.

Batschelet, E. (1974). *Introduction to Mathematics for Life Scientists*, Springer-Verlag, New York.

Beard, B. H. (1981). The Sunflower Crop, *Sci. Am.*, **244**, 150–161.

Beck, C. B., Schmid, R., and Rothwall, G. W. (1982). Stelar Morphology and the Primary Vascular System of Seed Plants, *Bot. Rev.*, **48**, 691–815, 913–931.

Bell, C. J. (1981). The Testing and Validation of Models, in *Mathematics and Plant Physiology*, D. A. Rose and D. A. Charles-Edwards, Eds., Academic Press, London, pp. 299–309.

Bellman, R. (1970). *Introduction to Matrix Analysis*, McGraw-Hill, New York.

Berding, C., Harbich, T., and Haken, H. (1983). A Pre-pattern Formation Mechanism for the Spiral-type Patterns of the Sunflower Head, *J. Theor. Biol.*, **104**, 53–70.

Berdyshev, A. P. (1972). On some Mathematical Regularities of Biological Processes, *Zh. Obshch. Biol.*, **33**, 631–638.

Berg, A. R., and Cutter, E. G. (1969). Leaf Initiation Rates and Volume Growth Rates in the Shoot Apex of *Chrysanthemum*, *Am. J. Bot.*, **56**, 153–159.

Berge, C. (1968). *Principes de Combinatoire*, Dunod, Paris.

Berge, C. (1976). *Graphs and Hypergraphs*, 2nd ed., Elsevier-North Holland, New York.

Bernal, J. D. (1965). Molecular Structure, Biochemical Function and Evolution, in *Theoretical and Mathematical Biology*, T. H. Waterman and H. J. Morowitz, Eds., Blaisdell, Boston, pp. 96–135.

Bertalanffy, L. von (1973). *Théorie Générale des Systèmes*, Dunod, Paris.

Bertalanffy, L. von (1974). The Model of Open Systems Beyond Molecular Biology, in *Biology, History and Natural Philosophy*, A. D. Breck and W. Yourgrau, Eds., Plenum-Rosetta, New York, pp. 17–30.

Beveridge, W. I. B. (1957). *The Art of Scientific Investigation*, Vintage, New York.

Bohm, D. (1969). Some Remarks on the Notion of Order, in *Towards a Theoretical Biology*, Vol. 2, *Sketches*, C. H. Waddington, Ed., Edinburgh University Press, pp. 18–60.

Bolle, F. (1939). Theorie der Blattstellung, *Verh. Bot. Ver. Prov. Brandenburg*, **79**, 152–192 (Berlin).

Bolle, F. (1963). Über Blattstellungsprobleme und Blütendurchwachsung bei der Nelke, *Ber. Dtsch. Bot. Ges.*, **76**, 211–228.

Bookstein, F. L. (1978). *The Measurement of Biological Shape and Shape Change*, Springer-Verlag, New York. [Reviewed by R. V. Jean, *Math. Biosci.*, **46**, 151–152 (1979).]

Bookstein, F. L. (1982). Foundations of Morphometrics, *Ann. Rev. Ecol. Syst.*, **13**, 451–470.

Braun, M. (1978). *Differential Equations and their Applications*, 2nd ed., Applied Mathematical Sciences Series, Vol. 15, Springer-Verlag, New York.

Bravais, L., and Bravais, A. (1837). Essai sur la Disposition des Feuilles Curvisériées, *Ann. Sci. Nat., Bot. Biol. Veg.*, **7**, 42–110, 193–221, 291–348; **8**, 11–42.

Bravais, L., and Bravais, A. (1839). Essai sur la Disposition Générale des Feuilles Rectisériées, *Ann. Sci. Nat. Bot.*, **12**, 5–14, 65–77.

Brillouin, L. (1949). Life, Thermodynamics and Cybernetics, *Am. Sci.*, **37**, 554–568.

Brillouin, L. (1959). *La Science et la Théorie de l'Information*, Masson, Paris.

Brousseau, A. (1968). On the Trail of the California Pine, *Fib. Q.*, **6**, 69–76.

Bruter, C. (1974a). Dynamique, Stabilité et Symétrie, *Rev. Bio.-Math.*, **45**, 7–8.

Bruter, C. (1974b). *Topologie et Perception*, Maloine-Drouin, Paris.

Buvat, R. (1955). Le Méristème Apical de la Tige, *Ann. Biol.*, **31**, 595–656.

Cahen, E. (1914). *Théorie des Nombres*, Vol. 1, Hermann, Paris.

Camefort, H. (1950). Anomalies Foliaires et Variations Phyllotaxiques chez les Plantules de *Cupressus sempervirens*, *Rev. Gen. Bot.*, **57**, 348–372.

Camefort, H. (1956). Etude de la Structure du Point Végétatif et des Variations Phyllotaxiques chez Quelques Gymnospermes, *Ann. Sci. Nat. Bot.*, **17**, 1–185.

Candolle, A. C. P. de (1881). *Considérations sur l'Etude de la Phyllotaxie*, Georg, Geneva.

Candolle, A. C. P. de (1895). Nouvelles Considérations sur la Phyllotaxie, *Arch. Sci. Phys. Nat.*, **33**, 121–147.

Carton, A. (1948). Etudes Phyllotaxiques sur Quelques Espèces de *Linum*, *Rev. Gen. Bot.*, **55**, 137–168.

Causton, D. R. (1969). A Computer Program for Fitting the Richards Function, *Biometrics*, **25**, 401–409.

Causton, D. R. (1977). *A Biologist's Mathematics*, Arnold, London.

Chadefaud, M. (1960). *Traité de Botanique*, Vol. 1, Masson, Paris.

Charles-Edwards, D. A., Cookshull, K. E., Horridge, J. S., and Thornley, (1979). A Model of Flowering in *Chrysanthemum*, *Ann. Bot.*, **44**, 557–566.

Cherruault, Y., Ed. (1982). *Recherches Biomathématiques: Méthodes et exemples*, Les cours du C.I.M.P.A., Nice.

Church, A. H. (1901). Note on Phyllotaxis, *Ann. Bot.*, **15**, 481–490.

Church, A. H. (1904a). *On the Relation of Phyllotaxis to Mechanical Laws*, William and Norgate, London.

Church, A. H. (1904b). The Principles of Phyllotaxis, *Ann. Bot.*, **18**, 227–243.

Church, A. H. (1920). *The Somatic Organization of the Phaeophyceae*, Oxford University Press, London.

Church, A. H. (1968). *On the Interpretation of Phenomena of Phyllotaxis*, Hafner, New York (facsimile of the 1920 edition).

Clark, L. B., and Hersh, A. H. (1939). A Study of Relative Growth, in *Notonecta undulata*, *Growth*, **3**, 347–372.

Clowes, F. (1961). *Apical Meristems*, Blackwell, Oxford.

Codaccionni, M. (1955). Etude Phyllotaxique d'un Lot de 200 Plants d'*Helianthus annuus* L. Cultivés en Serre, *C. R. Acad. Sci. Paris*, **241**, 1159–1161.

Collot, F., Boumokra, M., Casse, R., and Ricard, J. (1977). Concept d'une Quantité de p-Information Apportée par un Evènement Probable, *Rev. Bio.-Math.*, **58**.

Cook, T. A. (1914). *The Curves of Life*, Constable, London.

Corner, E. J. H. (1966). *The Natural History of Palms*, University of California Press, Berkeley.

Corner, E. J. H. (1981). *The Life of Plants*, The University of Chicago Press, Chicago.

Cotton, F. A. (1968). *Applications de la Théorie des Groupes à la Chimie*, Dunod, Paris.

Coxeter, H. S. M. (1969). *Introduction to Geometry*, Wiley, New York.

Coxeter, H. S. M. (1972). The Role of Intermediate Convergents in Tait's Explanation for Phyllotaxis, *J. Alg.*, **20**, 167–175.

Crafts, A. S. (1943a). Vascular Differentiation of the Shoot Apex of *Sequoia sempervirens*, *Am. J. Bot.*, **30**, 110–121.

Crafts, A. S. (1943b). Vascular Differentiation in the Shoot Apices of Ten Coniferous Species, *Am. J. Bot.*, **30**, 382–393.

Crank, J. (1976). *The Mathematics of Diffusion*, Clarendon Press, Oxford.

Crick, F. (1970). Diffusion in Embryogenesis, *Nature*, **225**, 420–422.

Crow, W. B. (1928). Symmetry in Organisms, *Am. Nat.*, **62**, 207–227.

Cuenod, A. (1951). Du Rôle de la Feuille dans l'Edification de la tige, *Bull. Soc. Sci. Nat. Tunisie*, **4**, 3–15.

Cull, P. and Vogt, A. (1973). Mathematical Analysis of the Asymptotic Behavior of the Leslie Population Matrix Model, *Bull. Math. Biol.*, **35**, 645–661.

Cutter, E. G. (1956). Experimental and Analytical Studies of Pteridophytes XXXIII. The Experimental Induction of Buds from Leaf Primordia in *Dryopteris aristata* Druce, *Ann. Bot. NS*, **20**, 143–168.

Cutter, E. G. (1959). On a Theory of Phyllotaxis and Histogenesis, *Biol. Rev.*, **34**, 243–263.

Cutter, E. G. (1964). Phyllotaxis and Apical Growth, *New Phytol.*, **63**, 39–46.

Cutter, E. G. (1965). Recent Experimental Studies on the Shoot Apex and Shoot Morphogenesis, *Bot. Rev.*, **31**, 7–113.

Cutter, E. G. (1966). Patterns of Organogenesis in the Shoot, in *Trends in Plant Morphogenesis*, E. G. Cutter, Ed., Longmans, London, pp. 220–234.

Cutter, E. G., and Voeller, B. R. (1959). Changes in Leaf Arrangement in Individual Fern Apices, *J. Linn. Bot.*, **56**, 225–236.

Davenport, H. (1952). *The Higher Arithmetic*, Hutchinson, London.

Davies, P. A. (1939). Leaf Position in *Ailanthus altissima* in Relation to the Fibonacci Series, *Am. J. Bot.*, **26**, 67–74.

Davis, T. A. (1963). The Dependence of Yield on Asymmetry in Coconut Palms, *J. Gen.*, **58**, 186–215.

Davis, T. A. (1964). Possible Geographical Influence on Asymmetry in Coconut and Other Plants, Tech. Working Paper No. IX on Coconut, Colombo, *Proc. FAO*, **2**, 59–69.

Davis, T. A. (1970). Fibonacci Numbers for Palm Foliar Spirals, *Acta. Bot. Neerl.*, **19**, 249–256.

Davis, T. A. (1971). Why Fibonacci Sequence for Palm Leaf Spirals, *Fib. Q.*, **9**, 237–244.

Davis, T. A., and Bose, T. K. (1971). Fibonacci Systems in Aroids, *Fib. Q.*, **9**, 253–263.

Davis, T. A., and Mathai, A. M. (1973). A Mathematical Explanation of the Emergence of Foliar Spirals in Palms, *Proc. Ind. Natl. Sci. Acad.*, **39**, 194–202.

Demetrius, L. (1981). The Malthusian Parameter and the Effective Rate of Increase, *J. Theor. Biol.*, **92**, 141–161.

Derome, J. R. (1977). Biological Similarity and Group Theory, *J. Theor. Biol.*, **65**, 369–378.

Deschatres, R. (1954). Recherches sur la Phyllotaxie du Genre *Sedum*, *Rev. Gen. Bot.*, **61**, 501–570.

Deysson, G. (1967). *Organisation et Classification des Plantes Vasculaires*, Vol. 2, Sedes, Paris.

Dixon, R. (1981). The Mathematical Daisy, *New Scientist*, **92**, 792–795.

Dormer, K. J. (1955). Mathematical Aspects of Plant Development, *Discovery*, **16**, 59–64.

Dormer, K. J. (1965). Correlations in Plant Development: General and Basic Aspects, in *Encyclopedia of Plant Physiology*, Vol. 1, W. Ruhland, Ed., Springer-Verlag, New York, pp. 451–491.

Dormer, K. J. (1972). *Shoot Organization in Vascular Plants*, Syracuse University Press, Syracuse.

Edwards, L. (1982). *The Field of Form*, Floris, Edinburgh.

Erickson, R. O. (1959). Patterns of Cell Growth and Differentiation in Plants, in *The Cell. Biochemistry, Physiology, Morphology*, Vol. 1, J. Brachet and A. E. Mirsky, Eds., Academic Press, New York, pp. 497–535.

Erickson, R. O. (1973). Tubular Packing of Spheres in Biological Fine Structure, *Science*, **181**, 705–716.

Erickson, R. O. (1976). Modeling of Plant Growth, *Ann. Rev. Plant Physiol.*, **27**, 407–434.

Erickson, R. O. (1983). The Geometry of Phyllotaxis, in *The Growth and Functioning of Leaves*, J. E. Dale and F. L. Milthorpe, Eds., Cambridge University Press, Cambridge, pp. 53–88.

Erickson, R. O., and Michelini, F. J. (1957). The Plastochron Index, *Am. J. Bot.*, **44**, 297–305.

Erickson, R. O. and Meicenheimer. R. D. (1977). Photoperiod Induced Change in Phyllotaxis in *Xanthium*, *Am. J. Bot.*, **64**, 981–988.

Esau, K. (1943). Vascular Differentiation in the Vegetative Shoot of *Linum*, II. The First Phloem and Xylem, *Am. J. Bot.*, **30**, 248–255.

Esau, K. (1954). Primary Vascular Differentiation in Plants, *Biol. Rev. Camb. Phil. Soc.*, **29**, 46–86.

Esau, K. (1965a). *Plant Anatomy*, 2nd ed., Wiley, New York.

Esau, K. (1965b). *Vascular Differentiation in Plants*, Holt-Rinehart-Winston, New York.

Esau, K. (1977). *Anatomy of Seed Plants*, 2nd ed., Wiley, New York.

Evans, G. C. (1972). *The Quantitative Analysis of Plant Growth*, Blackwell, Oxford.

Florin, R. (1951): Evolution in *Cordaites* and Conifers, *Acta Horti Bergiani*, **15**, 285–388.

Fomin, S. V., and Berkinblitt, M. B. (1975). *Problèmes Mathématiques en Biologie*, Mir, Moscow.

Fosket, E. B. (1968). The Relation of Age and Bud Break to the Determination of Phyllotaxis in *Catalpa speciosa*, *Am. J. Bot.*, **55**, 894–899.

Foster, A. S. (1939). Problems of Structure, Growth and Evolution in the Shoot Apex of Seed Plants, *Bot. Rev.* **5**, 454–470.

Fox, M. A. (1973). Some Thoughts on the Requirements for a Theory of Biology, *Bull. Math. Biol.*, **35**, 11–19.

Franke, W. H. (1967). Das ordnungsprinzip der Berührungszeilen, *Kakteen und andere Sukkulenten*, **18**, 207–209.

Franquin, P. (1974). Un Modèle Théorique du Développement de la Structure de la Plante, *Physiol. Veg.*, **12**, 459–465.

Frey-Wyssling, A. (1954). Divergence in Helical Polypeptide Chains and in Phyllotaxis, *Nature*, **173**, 596.

Frey-Wyssling, A. (1975). "Rechts" und "Links" im Pflanzenreich, *Biol. Unserer Zeit*, **5**, 147–154.

Frijters, D. (1978a). Principles of Simulation of Inflorescence Development, *Ann. Bot.*, **42**, 549–560.

Frijters, D. (1978b). Mechanisms of Developmental Integration of *Aster novae-angliae* L. and *Hieracium murorum* I., *Ann. Bot.*, **42**, 561–575.

Frijters, D., and Lindenmayer, A. (1976). Developmental Descriptions of Branching Patterns with Paracladial Relationships, in *Automata, Languages, Development*, A. Lindenmayer and G. Rozenberg, Eds., North-Holland, Amsterdam, pp. 57–73.

Fritsch, F. E. (1961). *The Structure and Reproduction of the Algae II*, Cambridge University Press, Cambridge.

Fujita, T. (1937). Über die Reihe 2,5,7,12,... in der Schraubigen Blattstellung und die Mathematische Betrachtung Verschiedener Zahlenreihensysteme, *Bot. Mag. Tokyo*, **51**, 480–489.

Fujita, T. (1938). Statistische Untersuchung über die Zahl der Konjugierten Parastichen bei den Schraubigen Organstellungen, *Bot. Mag. Tokyo*, **52**, 425–433.

Fujita, T. (1939). Statistische Untersuchungen über den Divergenzwinkel bei den Schraubigen Organstellungen, *Bot. Mag. Tokyo*, **53**, 194–199.

Fujita, T. (1942). Zurkenntnis der Organstellungen im Pflanzenreich, *Jap. J. Bot.*, **12**, 1–55.

Fujita, T. (1964). Phyllotaxis of *Cuscuta*, *Bot. Mag. Tokyo*, **77**, 73–76.

Gantmacher, F. R. (1959). *Theory of Matrices*, Vols. 1 and 2, Chelsea, New York.

Gauquelin, M. (1973). *Rythmes Biologiques, Rythmes Cosmiques*, Marabout Université, Verviens.

Gifford, E. M., and Corson, C.E. (1971). The Shoot Apex in Seed Plants, *Bot. Rev.*, (Lancaster), **37**, 143–229.

Gifford, E. M., and Tepper, H. B. (1962). Ontogenetic and Histochemical Changes in the Vegetative Shoot Tip of *Chenopodium album*, *Am. J. Bot.*, **49**, 902–911.

Girolami, G. (1953). Relation between Phyllotaxis and Primary Vascular Organization in *Linum*, *Am. J. Bot.*, **40**, 618–625.

Glansdorff, P., and Prigogine, I. (1971). *Structure, Stabilité et Fluctuation*, Masson, Paris.

Goebel, K. (1969). *Organography of Plants*, Hafner, New York (facsimile of the 1900 edition).

Goldberg, S. (1961). *Introduction to Difference Equations with Illustrative Examples from Economics, Psychology and Sociology*, Wiley, New York.

Gomez-Campo, C. (1974). Phyllotactic Patterns in *Bryophyllum Tubiflorum* Harv., *Bot. Gaz*, **135**, 49–58.

Grobstein, C. (1962). Levels and Ontogeny, *Am. Sci.*, **50**, 46–58.

Guédès, M., and Dupuy, P. (1983). From Decussation to Distichy, with some Comments on Current Theories of Phyllotaxis, *Bot. J. Linn. Soc.*, **87**, 1–12.

Guiaşu, S. (1977). *Information Theory with Applications*, McGraw-Hill, New York.

Guinochet, M. (1965). *Notions Fondamentales de Botanique Générale*, Masson, Paris.

Gunckel, J. E. (1965). Modifications of Plant Growth and Development induced by Ionizing Radiations, in *Encyclopedia of Plant Physiology*, Vol. 2, W. Ruhland, Ed., Springer-Verlag; New York, pp. 365–387.

Gunckel, J. E., and Wetmore, R. H. (1946). Studies of Development in Long Shoots and Short Shoots of *Ginkgo biloba* L. II. Phyllotaxis and the Organization of the Primary Vascular System; Primary Phloem and Primary Xylem, *Am. J. Bot.*, **33**, 532–543.

Günther, B., and Guerra, E. (1955). Biological Similarities, *Acta Physiol. Latino am.*, **5**, 169–186.

Halberg, C. J. A. (1955). *Spectral Theory of Linked Operators in the l^p Spaces*, Doctoral Thesis, UCLA.

Hallé, N. (1979a). Analyse du Réseau Phyllotaxique des Ecussons du Cône chez *Pinus*, *Adansonia*, ser. 2, **18** (4), 393–408.

Hallé, N. (1979b). Essai de Phyllotaxie Dynamique Interprétée comme Primitive, *Bull. Mus. Natl. Hist. Nat. Paris*, 4e Ser, 1, Section B, No. 2, 71–95.

Hallé, N. (1979c). Sur une nouvelle méthode descriptive du Réseau Phyllotaxique des Ecussons du Cône chez *Pinus*, L. et son intérêt taxonomique, *C. R. Acad. Sc. Paris*, ser. D, **288**, 59–62.

Halperin, W. (1978). Organogenesis at the Shoot Apex, *Ann. Rev. Plant Physiol.*, **29**, 239–262.

Hamermesh, M. (1962). *Group Theory and its Application to Physical Problems*, Addison-Wesley, London.

Hancock, H. (1964). *Development of the Minkowski Geometry of Numbers*, Dover, New York.

Harary, F. (1955). The Number of Linear, Directed, Rooted and Connected Graphs, *Trans. Am. Math. Soc.*, **78**, 445–463.

Harary, F. (1960). Unsolved Problems in the Enumeration of Graphs, *Publ. Math. Inst. Hung. Acad. Sci.*, **5**, 63–95.

Hardy, G. H., and Wright, E. M. (1960). *An Introduction to the Theory of Numbers*, 4th ed., Clarendon Press, Oxford.

Harris, T. E. (1963). *The Theory of Branching Processes*, Springer-Verlag, Berlin.

Harris, W. F., and Erickson, R. O. (1980). Tubular Arrays of Spheres: Geometry, Continuous and Discontinuous Contraction, and the Role of Moving Dislocations, *J. Theor. Biol.*, **83**, 215–246.

Haynes, K. E. (1980). The Entropies: Some Roots of Ambiguity, *Socio-Econ. Plan. Sci.*, **14**, 137–145.

Heimans, J. (1978). Problems of Phyllotaxis, *Proc. Kon. Nederl. Akad. Wet.*, **81C**, 91–98.

Hellendoorn, P. H., and Lindenmayer, A. (1974). Phyllotaxis in *Bryophyllum tubiflorum*: Morphogenetic Studies and Computer Simulations, *Acta. Bot. Neerl.*, **23**, 473–492.

Heller, R. (1978). *Abregé de Physiologie Végétale*, Masson, Paris.

Herman, G. T., Lindenmayer, A., and Rozenberg, G. (1975). Description of Developmental Languages Using Recurrence Systems, *Math. Syst. Theory*, **8**, 316–341.

Herman, G. T., and Rozenberg, G. (1975). *Developmental Systems and Languages*, North Holland, Amsterdam and American Elsevier, New York.

Herman, G. T., and Vitanyi, P. M. B. (1976). Growth Functions Associated with Biological Development, *Am. Math. Mon.*, **83**, 1–16.

Herschkowitz-Kaufman, M. (1977). Structures Dissipatives dans les Systèmes Chimiques Hors d'Equilibre, *Rev. Bio.-Math.*, **60**, 235–251.

Hersh, A. H. (1941). Allometric Growth: the Ontogenic and Phylogenic Significance of Differential Rates of Growth, *Growth* (suppl.), **5**, 113–145.

Hersh, A. H., and Anderson, B. G. (1941). Differential Growth and Morphological Pattern in *Daphnia*, *Growth*, **5**, 359–364.

Hirmer, M. (1922). *Zur Lösung des Problems der Blattstellungen*, Fisher, Jena.

Hofmeister, W. (1868). Allgemeine Morphologie der Gewachse, *Handbuch der Physiologischen Botanik*, 1, Engelmann, Leipzig, pp. 405–664.

Holland, J. M. (1972). *Studies in Structure*, Macmillan, New York.

Hunt, R. (1981). The Fitted Curve in Plant Growth Studies, in *Mathematics and Plant Physiology*, D. A. Rose and D. A. Charles-Edwards, Eds., Academic Press, London, pp. 283–298.

Huntley, H. E. (1970). *The Divine Proportion. A Study in Mathematical Beauty*, Dover, New York.

Huxley, J. S. (1932, issued in 1972). *Problems of Relative Growth*, Dover, New York.

Iterson, G. van (1907). *Mathematische and Mikroskopisch-Anatomische Studien über Blattstellungen, nebst Betrschtungen über den Schalenbau der Miliolinen*, Gustav-Fisher-Verlag, Jena.

Iterson, G. van (1960). New Studies on Phyllotaxis, *Proc. Kon. Nederl. Akad. Wet.*, **63C**, 137–150.

Jean, R. V. (1976). Growth Matrices in Phyllotaxis, *Math. Biosci.*, **32**, 65–76.

Jean, R. V. (1978a). Growth and Entropy: Phylogenism in Phyllotaxis, *J. Theor. Biol.*, **71**, 639–660.

Jean, R. V. (1978b). *Phytomathématique*, Les Presses de l'Université du Québec, Québec. [Reviewed by G. Estabrook, *Math. Biosci.*, **46**, 301–302 (1979).]

Jean, R. V. (1979a). Developmental Algorithms in Phyllotaxis and their Implications: a Contribution to the Field of Growth Functions of L-Systems by an Application of Perron-Frobenius Spectral Theory, *J. Theor. Biol.*, **76**, 1–30.

Jean, R. V. (1979b). A Rigorous Treatment of Richards' Mathematical Theory of Phyllotaxis, *Math. Biosci.*, **44**, 221–240.

Jean, R. V. (1980a). La Notion Systémique de Contenu d'Information Structurelle et ses Dérivés, *Rev. Bio.-Math.*, **70**, 5–25.

Jean, R. V. (1980b). Contribution à la Théorie des Fonctions de Croissance des L-Systèmes et à la Biométrie Végétale, *C. R. Acad. Sci. Paris, Sect. Anal. Math.*, **290A**, 949–952.

Jean, R. V. (1980c). A Systemic Model of Growth in Botanometry, *J. Theor. Biol.*, **87**, 569–584.

Jean, R. V. (1981a). An L-System Approach to Non Negative Matrices for the Spectral Analysis of Discrete Growth Functions of Populations, *Math. Biosci.*, **55**, 155–168.

Jean, R. V. (1981b). The Use of Continued Fractions in Botany, *UMAP Module Unit* **U571**, Lexington, Mass.

Jean, R. V. (1981c). A New Approach to a Problem of Plant Growth, in *Applied Systems and Cybernetics*, G. E. Lasker, Ed., Pergamon Press, New York, pp. 1906–1910.

Jean, R. V. (1982). The Hierarchical Control of Phyllotaxis, *Ann. Bot.*, **49**, 747–760.

Jean, R. V. (1983a). *Croissance végétale et Morphogénèse*, Masson, Paris, and Presses de l'Université du Québec, Quebec.

Jean, R. V. (1983b). Mathematical Modeling in Phyllotaxis: the State of the Art, *Math. Biosci.*, **64**, 1–27.

Jean, R. V. (1983c). Le Phénomène de la Phyllotaxie des Plantes, *Ann. Sci. Nat. Bot.*, **13**, 5, 45–46.

Jean, R. V. (1983d). The Fibonacci Sequence, *UMAP Journal*,[5] in press.

Jean, R. V. (1983e). Differential Growth, Huxley's Allometric Formula, and Sigmoid Growth, *UMAP Module Unit* **U635**, Lexington, Mass.

Jean, R. V. (1983f). A Fundamental Problem in Plant Morphogenesis, from the Standpoint of Differential Growth, in *Mathematical Modelling in Science and Technology*, X. J. R. Avula, Ed., Pergamon Press, New York, pp. 774–777.

Jean, R. V. (1983g). Allometric Relations in Plant Growth, *J. Math. Biol.*, **18**, 189–200.

Jensen, L. C. W. (1968). Primary Stem Vascular Patterns in Three Subfamilies of the Crassulaceae, *Am. J. Bot.*, **55**, 553–563.

Katz, M. B. (1978). *Questions of Uniqueness and Resolution in Reconstruction from Projections*, Lecture Notes in Biomathematics, Springer-Verlag, New York. [Reviewed by R. V. Jean, *Math. Biosci.*, 49, 155–156 (1980).]

Kavanagh, A. J., and Richards, O. W. (1942). Mathematical Analysis of the Relative Growth of Organisms, *Proc. Rochester Acad. Sci.*, **8**, 150–174.

Kerns, K. R., Collins, J. L., and Kim, H. (1936). Developmental Studies of the Pineapple, *Ananas comosus* (L.) Merr. I. Origin and Growth of Leaves and Inflorescence, *New Phytol.*, **35**, 305–317.

Khintchine, A. I. (1957). *Mathematical Foundations of Information Theory*, Dover, New York.

Khintchine, A. I. (1963). *Continued Fractions*, Noordhoffeltd, Groningen.

Kilmer, W. L. (1971). On Growing Pine Cones and Other Fibonacci Fruits; McCulloch's Localized Algorithm, *Math. Biosci.*, **11**, 53–57.

King, C. E., Gallaher, E. E., and Levin, D. A. (1975). Equilibrium Diversity in Plant Pollinator Systems, *J. Theor. Biol.*, **53**, 263–275.

Kitagawa, T. (1971). A Contribution to the Methodology of Biomathematics: Information Science Approach to Biomathematics I, *Math. Biosci.*, **2**, 329–345.

Klein, F. (1907). *Ausgewählte Kapitel der Zahlentheorie*, Teubner, Gottingen.

Klein, F. (1932). *Elementary Mathematics from an Advanced Standpoint*, Macmillan, New York.

Kofler, L. (1963). *Croissance et Développement des Plantes*, Gauthier-Villars, Paris.

Kruger, F. (1973). On the Problem of Summing up the Allometric Formula, *Rev. Bio.-Math.*, **12**, 123–128.

Kuich, W. (1970). On the Entropy of Context-Free Languages, *Inf. Control*, **16**, 173–200.

Larson, P. R. (1977). Phyllotactic Transitions in the Vascular System of *Populus deltoïdes* Barts, as determined by ^{14}C labelling, *Planta*, **134**, 241–249.

Larson, P. R. (1983). Primary Vascularization and the Siting of Primordia, in *The Growth and Functioning of Leaves*, J. E. Dale and F. L. Milthorpe, Eds., Cambridge University Press, Cambridge, pp. 25–51.

Lehninger, A. L. (1969). *Bioénergétique*, Ediscience, Paris.

Leigh, E. G. (1972). The Golden Section and Spiral Leaf-Arrangement, *Trans. Conn. Acad. Arts Sci.*, **44**, 163–176.

Leppik, E. E. (1961). Phyllotaxis, Anthotaxis and Semataxis, *Acta Biotheor.*, **14**, 1–28.

Leppik, E. E. (1970). Evolutionary Differentiation of the Flower Head of the Compositae II, *Ann. Bot. Fennici*, **7**, 325–352.

Leveque, W. S. (1956). *Topics in Number Theory*, Vol. 1, Addison-Wesley, Reading, Mass.

Lindenmayer, A. (1971). Developmental Systems without Cellular Interaction: their Languages and Grammars, *J. Theor. Biol.*, **30**, 455–484.

Lindenmayer, A. (1975). Developmental Algorithms for Multicellular Organisms: a Survey of L-systems, *J. Theor. Biol.*, **54**, 3–22.

Lindenmayer, A. (1977a). Theories and Observations of Developmental Biology, in *Foundational Problems in Special Sciences*, Butts and Hintikka, Eds., Reidel, Dordrecht, Holland, pp. 103–118.

Lindenmayer, A. (1977b). Paracladial Relationships in Leaves, *Ber. Dtsch. Bot. Ges. Bd.*, **90**, 287–301.

Lindenmayer, A. (1978a). Algorithms for Plant Morphogenesis, in *Theoretical Plant Morphology*, R. Sattler, Ed., Leiden University Press, The Hague, pp. 37–81 (suppl. to *Acta Biotheoretica*, 27).

Lindenmayer, A. (1978b). Growth Functions of Cell Populations with Lineage Control, in *Biomathematics and Cell Kinetics* A. J. Valleron and P. D. M. MacDonald, Eds., Elsevier, New York, pp. 117–133.

Loiseau, J. E. (1957). Evolution de la Phyllotaxie chez *Tropaeolum majus* L. Cultivé en Chambres Lumineuses Conditionnées, *Bull. Sc. Bourgogne*, **18**, 57–59.

Loiseau, J. E. (1959). Observations et Expérimentations sur la Phyllotaxie et le Fonctionnement du Sommet Végétatif chez Quelques Balsaminacées, *Ann. Sci. Nat. Bot.*, **20**, 1–214.

Loiseau, J. E. (1969). *La Phyllotaxie*, Masson, Paris.

Loiseau, J. E., and Deschatres, R. (1961). Les Phyllotaxies Bijuguées, *Mem. Bull. Soc. Bot. Fr.*, **108**, 105–116.

Loiseau, J. E., and Messadi, M. (1957). Observations sur des Phyllotaxies à angles oscillants, *Bull. Sci. Bourgogne*, **18**, 61–64.

Lovtrup, S., and Sydow, B. von (1974). D'Arcy Thompson's Theorems and the Shape of the Molluscan Shell, *Bull. Math. Biol.*, **36**, 567–575.

Lück, H. B. (1975). Elementary Behavioural Rules as a Foundation for Morphogenesis, *J. Theor. Biol.*, **54**, 23–34.

Lyndon, R. F. (1972). Leaf Formation and Growth at the Shoot Apical Meristem, *Physiol. Veg.*, **10**, 209–222.

Lyndon, R. F. (1978). Phyllotaxis and the Initiation of Primordia During Flower Development in *Silene*, *Ann. Bot.*, **42**, 1349–1360.

Lyndon, R. F. (1979). A Modification of Flowering and Phyllotaxis in *Silene*, *Ann. Bot.*, **43**, 553–558.

McCully, M. E., and Dale, H. M. (1961). Variations in Leaf Number in *Hippuris*. A Study of Whorled Phyllotaxis, *Can. J. Bot.*, **39**, 611–625.

Magnus, W., and Grossman, I. (1971). *Les Groupes et leurs Graphes*, Dunod, Paris.

Mahler, K., Cassels, J. W. S., and Ledermann, W. (1951). Farey Section in $k(i)$ and $k(p)$, *Phil. Trans. Roy. Soc. London*, **243A**, 585–628.

Majumdar, G. P. (1948). Leaf Development at the Growing Apex and Phyllotaxis in *Heracleum*, *Proc. Ind. Acad. Sci.*, **28**, 83–98.

Maksymowych, R. (1973). *Analysis of Leaf Development*, Cambridge University Press, Cambridge.

Maksymowych, R., and Erickson, R. O. (1977). Phyllotactic Change Induced by Gibberellic Acid on Xanthium Shoot Apices, *Am. J. Bot.*, **64**, 33–44.

Marzec, C., and Kappraff, J. (1983). Properties of Maximal Spacing on a Circle Related to Phyllotaxis and to the Golden Mean, *J. Theor. Biol.*, **103**, 201–226.

Mathai, A. M., and Davis, T. A. (1974). Constructing the Sunflower Head, *Math. Biosci.*, **20**, 117–133.

Maynard-Smith, J. (1971). *Mathematical Ideas in Biology*, Cambridge University Press, Cambridge.

Meicenheimer, R. D. (1979). Relationships between Shoot Growth and Changing Phyllotaxis of *Ranunculus*, *Am. J. Bot.*, **66**, 557–569.

Meicenheimer, R. D. (1980). Growth Characteristics of *Epilobium hirsutum* Shoots Exhibiting Bijugate and Spiral Phyllotaxy, Ph.D Dissertation, Department of Botany, Washington State University.

Meicenheimer, R. D. (1981). Changes in *Epilobium* Phyllotaxy Induced by N-1-Naphthyl-phthalamic Acid and α-4-Chlorophenoxyisobutyric Acid, *Am. J. Bot.*, **68**, 1139–1154.

Meinhardt, H. (1974). The Formation of Morphogenetic Gradients and Fields, *Ber. Dtsch. Bot. Ges. Bd.*, **87**, 101–108.

Meinhardt, H., and Gierer, A. (1974). Applications of a Theory of Biological Pattern Formation Based on Lateral Inhibition, *J. Cell. Sci.*, **15**, 321–346.

Mesarovic, M. D. (1968). Systems Theory and Biology—View of a Theoretician, in *Systems Theory and Biology*, M. D. Mesarovic, Ed., Springer-Verlag, New York, pp. 59–87.

Mesarovic, M. D. (1972). A Mathematical Theory of General Systems, in *Trends in General Systems Theory*, G. Klir, Ed., Wiley-Interscience, New York, pp. 251–269.

Michelini, F. J. (1958). The Plastochron Index in Developmental Studies of *Xanthium italicum* Moretti, *Am. J. Bot.*, **45**, 525–533.

Millener, L. H. (1952). An Experimental Demonstration of the Dependence of Phyllotaxis on Rate of Growth, *Nature*, **169**, 1052–1053.

Milthorpe, F. L. Ed., (1956). *The Growth of Leaves*, Butterworths, London.

Mitchison, G. H. (1977). Phyllotaxis and the Fibonacci Series, *Science*, **196**, 270–275.

Mohr, H. (1982). Principles in Plant Morphogenesis, in *Axioms and Principles of Plant Construction*, R. Sattler, Ed., Nyhoff and Junk, Boston, pp. 93–111.

Morowitz, H. J. (1972). *Entropy for Biologists—an Introduction to Thermodynamics*, 3rd ed., Academic Press, New York.

Mowshowitz, A. (1967). *Entropy and the Complexity of Graphs*, Technical Report Concomp, The University of Michigan. [Doctoral Thesis which appeared in four articles in *Bull. Math. Biophys.*, **30** (1968).]

Nägeli, C. (1858). Das Wachstum des Stammes und der Wurzel bei den Gefässpflanzen und die Anordnung der Gefässtränge im Stengel, *Beitr. Wiss. Bot.*, **1**, 1–156 (Engelmann, Leipzig).

Namboodiri, K. K., Beck, C. B. (1968). A Comparative Study of the Primary Vascular System of Conifers, I, II, III, *Am. J. Bot.*, **55**, 447–472.

Nicolis, G., and I. Prigogine (1977). *Self-Organization in Non-Equilibrium Systems*, Wiley, New York.

Northrop, E. P. (1975). *Riddles in Mathematics, a Book in Paradoxes*, Krieger, New York.

O'Connell, D. T. (1951). *Sci. Mon.*, **73**, 333.

Olds, C. D. (1963). *Continued Fractions*, Random House, Toronto.

O'Neil, T. (1961). Primary Vascular Organization of *Lupinus* Shoot, *Bot. Gaz.*, **123**, 1–9.

Papentin, F. (1980). On Order and Complexity. I. General Considerations, *J. Theor. Biol.*, **87**, 421–456.

Pattee, H. H. (1970). The Problem of Biological Hierarchy, in *Towards a Theoretical Biology*, Vol. 3, Drafts, C. H. Waddington, Ed., Edinburgh University Press, Edinburgh, pp. 117–136.

Pattee, H. H. Ed. (1973). *Hierarchy Theory, the Challenge of Complex Systems*, Braziller, New York.

Pauling, L., and Corey, R. B. (1953). Compound Helical Configurations of Polypeptide Chains: Structure of Proteins of the α-keratin Type, *Nature*, **171**, 59.

Paz, A., and Salomaa, A. (1973). Integral Sequential Word Functions and Growth Equivalence of Lindenmayer Systems, *Inf. Control*, **23**, 313–343.

Peusner, L. (1974). *Concepts in Bioenergetics*, Prentice-Hall, Englewood Cliffs, N.J.

Philipson, W. R. (1949). The Ontogeny of the Shoot Apex in Dicotyledons, *Biol. Rev.*, **24**, 21–50.

Philipson, W. R., and Balfour, E. E. (1963). Vascular Patterns in Dicotyledons, *Bot. Rev.*, **29**, 382–404.

Piélou, E. C. (1969). *Introduction to Mathematical Ecology*, Wiley-Interscience, New York.

Pilet, P. E. (1967). *L'Energétique Végétale*, les Presses Universitaires de France, Paris.

Pippenger, N. (1978). La Théorie de la Complexité, *Pour la Science*, **10**, 86–95.

Plantefol, L. (1946). Sur les Méthodes en Phyllotaxie, *C. R. Acad. Sci. Paris*, **222**, 1508–1510.

Plantefol, L. (1948). *La Théorie des Hélices Foliaires Multiples*, Masson, Paris.

Plantefol, L. (1950). La Phyllotaxie, *Année Biol.*, **54**, 447–460.

Plantefol, L. (1951). Phyllotaxie et Point Végétatif, *Scientia*, **86**, 91–98.

Plantefol, L. (1956). Sur les Variations Phyllotaxiques du *Stapelia hirsuta*, *C. R. Acad. Sci. Paris*, **243**, 916–919.

Pollard, J. H. (1973). *Mathematical Models for Growth of Human Populations*, Cambridge University Press, Cambridge.

Popper, K. (1968). *The Logic of Scientific Discovery*, Harper-Row, New York.

Priestley, J. H., and Scott, L. I. (1936). The Vascular Anatomy of *Helianthus annus* L., *Proc. Leeds Phil. Lit. Soc.*, **3**, 159–173.

Priestley, J. H., Scott, L. I., and Gillett, E. C. (1935). The Development of the Shoot in *Alstroemeria* and the Unit of Shoot Growth in Monocotyledons, *Ann. Bot.*, **49**, 161–179.

Priestley, J. H., Scott, L. I., and Mattinson, K. M. (1937). Dicotyledon Phyllotaxis from the Standpoint of Development, *Proc. Leeds Phil. Lit. Soc.*, **3**, 380–388.

Prigogine, I., and Glansdorff, P. (1971). *Structure, Stabilité et Fluctuation*, Masson, Paris.

Rashevsky, N. (1955). Life, Information Theory and Topology, *Bull. Math. Biophys.*, **17**, 229–235.

Rashevsky, N. (1960). *Mathematical Biophysics, Physico-Mathematical Foundations of Biology*, Vols. 1 and 2, Dover, New York.

Rashevsky, N. (1961). *Mathematical Principles in Biology and their Applications*, Thomas, Springfield.

Rashevsky, N. (1965). Models and Mathematical Principles in Biology, in *Theoretical and Mathematical Biology*, T. H. Watermann and H. J. Morowitz, Eds. Blaidel, Toronto, pp. 36–54.

Rashevsky, N. (1968). Organismic sets II; Some General Considerations, *Bull. Math. Biophys.*, **30**, 163–174.

Rees, A. R. (1964). The Apical Organization and Phyllotaxis of the Oil Palm, *Ann. Bot. N.S.*, **28**, 57–69.

Reeve, E. C. R., and Huxley, J. S. (1972). Some Problems in the Study of Allometric Growth, in *Problems of Relative Growth*, J. S. Huxley, Ed., Dover; New York, pp. 267–303.

Reinberg, A. (1977). *Des Rhythmes Biologiques à la Chronobiologie*, Gauthier-Villars, Paris.

Rényi, A. (1961). On Measures of Entropy and Information, *Proceedings Fourth Berkeley Symposium on Mathematical and Statistical Problems*, Vol. 1, University of California Press, Berkeley, pp. 547–561.

Ricard, J. (1960). *La Croissance des Végétaux*, Que sais-je? Presses Univ. de France, Paris.

Ricciardi, L. M. (1977). *Diffusion Processes and Related Topics*, (in the Series Lecture Notes in Biomathematics, **14**), Springer-Verlag, New York.

Richards, F. J. (1948). The Geometry of Phyllotaxis and its Origin, *Symp. Soc. Exp. Biol.*, **2**, 217–245.

Richards, F. J. (1951). Phyllotaxis: its Quantitative Expression and Relation to Growth in the Apex, *Phil. Trans. R. Soc. London*, **235B**, 509–564.

Richards, F. J. (1955). Über Einige Fragen Betreffend die Messung von Blattstellungen, *Planta*, **45**, 198–207.

Richards, F. J. (1956). Spatial and Temporal Correlations Involved in Leaf Pattern Production at the Apex, in *The Growth of Leaves*, F. L. Milthorpe, Ed., Butterworths, London, pp. 66–76.

Richards, F. J. (1959). A Flexible Growth Function for Empirical Use, *J. Exp. Bot.*, **10**, 290–300.

Richards, F. J. (1969). The Quantitative Analysis of Growth, in *Plant Physiology a Treatise*, Vol. 5A, F. C. Steward, Ed., Academic Press, New York, pp. 3–76.

Richards, F. J., and Schwabe, W. W. (1969). Phyllotaxis: a Problem of Growth and Form, in *Plant Physiology a Treatise*, Vol. 5A, F. C. Steward, Ed., Academic Press, New York, pp. 79–116.

Richards, O. W., and Kavanagh, A. J. (1945). The Analysis of Growing Form, in *Essays on Growth and Form*, W. E. Le Gros Clark and P. B. Medawar, Eds., Oxford University Press, London and New York, pp. 188–230.

Richter, P. H., and Schranner, R. (1978). Leaf Arrangement—Geometry, Morphogenesis and Classification, *Naturwissenschaften*, **65**, 319–327.

Ridley, J. N. (1982a). Packing Efficiency in Sunflower Heads, *Math. Biosci.*, **58**, 129–139.

Ridley, J. N. (1982b). Computer Simulation of Contact Pressure in Capitula, *J. Theor. Biol.*, **95**, 1–11.

Rijven, A. H. G. C. (1969). Randomness in the Genesis of Phyllotaxis II, *New Phytol.*, **68**, 377–386.

Roberts, D. W. (1977). A Contact Pressure Model for Semidecussate and Related Phyllotaxis, *J. Theor. Biol.*, **68**, 583–597.

Roberts, D. W. (1978). The Origin of Fibonacci Phyllotaxis—An Analysis of Adler's Contact Pressure Model and Mitchison's Expanding Apex Model, *J. Theor. Biol.*, **74**, 217–233.

Rosen, R. (1960). A Quantum—Theoretic Approach to Genetic Problems, *Bull. Math. Biophys.*, **22**, 227–255.

Rosen, R. (1967). *Optimality Principles in Biology*, Butterworths, London.

Rosen, R. (1972). *Foundations of Mathematical Biology*, Academic Press, New York.

Rosen, R. (1978). *Fundamentals of Measurement and Representation of Natural Systems*, North-Holland, New York.

Rozenberg, G. (1976). Bibliography of L-Systems, in *Automata, Languages, Development*, A. Lindenmayer and G. Rozenberg, Eds., North-Holland, Amsterdam, pp. 351–366.

Rozenberg, G., and Lindenmayer, A. (1973). Developmental Systems with Locally Catenative Formulas, *Acta Informatica*, **2**, 214–248.

Rutishauser, R. (1981). *Blattstellung und Sprossentwicklung bei Blütenpflanzen*, Dissertationes Botanicae, Vol. 62, Cramer, Vaduz.

Rutishauser, R. (1982). Der Plastochronquotient als Teil einer quantitativen Blattstellungsanalyse bei Samenpflanzen, *Beitr. Biol. Pflanzen*, **57**, 323–357.

Sachs, J. (1882). *Text-Book of Botany*, 2nd ed., Oxford University Press, London and New York.

Salomaa, A., and Soittola, M. (1978). *Automata-Theoretic Aspects of Formal Power Series*, Springer, New York.

Sattler, R. (1974). A New Conception of Shoot of Higher Plants, *J. Theor. Biol.*, **47**, 367–382.

Sattler, R. (1978). What is Theoretical Plant Morphology?, in *Theoretical Plant Morphology*, R. Sattler, Ed., Leiden Univ. Press, The Hague, pp. 5–20 (suppl. to *Acta Biotheoretica*, **27**).

Schimper, C. F. (1836). Geometrische Anordnung der um eine Axe Periferischen Blattgebilde, *Verhandl. Schweiz. Naturf. Ges.*, **21**, 113–117.

Schoute, J. C. (1913). Beiträge zur Blattstellungslehre, I. Die Theorie, *Rec. Trav. Bot. Neer.*, **10**, 153–339.

Schoute, J. C. (1936). On Whorled Phyllotaxis, *Rec. Trav. Bot. Neer.*, **33**, 649–669, 670–687.

Schrodinger, E. (1962). *What is Life?* Cambridge University Press, Cambridge.

Schuepp, O. (1946). Geometrische Betrachtungen über Wachstum und Formwechsel, *Ber. Schweiz. Bot. Ges.*, **56**, 629–655.

Schuepp, O. (1959). Konstruktionen zur Theorie der Blattstellung, *Denkschr. Schweiz. Naturf. Ges.*, **82**, 103–197.

Schuepp, O. (1963). Mathematisches und Botanisches über Allometrie, *Verhandl. Naturf. Ges. Basel*, **74**, 69–105.

Schuepp, O. (1966). *Meristeme*, Birkhäuser-Verlag, Basel, Stuttgart.

Schwabe. W. W. (1971). Chemical Modification of Phyllotaxis and its Implications, in *Control Mechanisms of Growth and Differentiation*, D. C. Davies and M. Balls, Eds., *Symp. Soc. Exp. Biol.*, **25**, Cambridge University Press, Cambridge, pp. 301–322.

Schwendener, S. (1878). *Mechanische Theorie der Blattstellungen*, Engelmann, Leipzig.

Schwendener, S. (1909). *Theorie der Blattstellungen, Mechanische Probleme der Botanik*, Engelmann, Leipzig.

Shannon, C. E. (1948). A Mathematical Theory of Communication, *Bell Syst. Tech. J.*, **27**, 379–656.

Shannon, C. E., and Weaver, N. (1967). *The Mathematical Theory of Communication*, 11th ed., The University of Illinois Press, Urbana.

Sharp, W. E. (1971). An Analysis of the Laws of Stream Order for Fibonacci Drainage Patterns, *Water Res.*, **7**, 1548–1557.

Shool, D. A. (1954). Regularities in Growth Curves, Including Rhythm and Allometry, in *Dynamics of Growth Processes*, J. E. Boell, Ed., Princeton University Press, Princeton, N.J., 224–241.

Sifton, H. V. (1944). Developmental Morphology of Vascular Plants, *New Phytol.*, **43**, 87–129.

Silk, K. W., and Erickson, R. O. (1979). Kinematics of Plant Growth, *J. Theor. Biol.*, **76**, 481–501.

Simon, H. A. (1962). The Architecture of Complexity, *Proc. Am. Phil. Soc.*, **106**, 467–482.

Simon, H. A. (1977). *Models of Discovery*, Reidel, Boston.

Simpson, G. G. (1967). *The Major Features of Evolution*, Columbia University Press, New York.

Sinnott, E. W. (1960). *Plant Morphogenesis*, McGraw-Hill, New York.

Sinnott, E. W. (1966). The Geometry of Life, in *Trends in Plant Morphogenesis*, E. G. Cutter, Ed., Longmans, London, 88–93.

Skipworth, J. P. (1962). The Primary Vascular System and Phyllotaxis in *Hectorella caespitosa* Hook. f., *N. Z. J. Sci.*, **5**, 253–258.

Slover, J. de (1958). Le Sens Longitudinal de la Differentiation du Procambium, du Xylène et du Phloème chez *Coleus, Ligustrum, Anagallis* et *Taxus*, *La Cellule*, **59**, 55–202.

Smith, B. W. (1941). The Phyllotaxis of *Costus* from the Standpoint of Development, *Proc. Leeds Phil. Soc.*, **4**, 42–63.

Smith, R. (1980). Rethinking Allometry, *J. Theor. Biol.*, **87**, 97–111.

Snow, M. (1955). Spirodistichy Re-interpreted, *Phil. Trans. Roy. Soc. London*, **239B**, 45–88.

Snow, M., and R. Snow (1931). Experiments on Phyllotaxis. I. The Effect of Isolating a Primordium, *Phil. Trans. Roy. Soc. London*, **221B**, 1–43.

Snow, M., and R. Snow (1934). The Interpretation of Phyllotaxis, *Biol. Rev., Cambridge Phil. Soc.*, **9**, 132–137.

Snow, M., and R. Snow (1935). Experiments on Phyllotaxis. III. Diagonal Splits Through Decussate Apices, *Phil. Trans. Roy. Soc. London*, **225B**, 63–94.

Snow, M., and R. Snow (1937). Auxin and Leaf Formation, *New Phytol.*, **36**, 1–18.

Snow, M., and Snow, R. (1947–1948). On the Determination of Leaves, *New Phytol.*, **46**, 5–19; *Soc. Exp. Biol. Symp.*, **2**, 263–275.

Snow, M., and Snow, R. (1952). Minimum Areas and Leaf Determination, *Proc. R. Soc. London*, **139B**, 545–566.

Snow, M., and Snow, R. (1962). A Theory of the Regulation of Phyllotaxis Based on *Lupinus albus*, *Phil. Trans. Roy. Soc. London*, **244B**, 483–513.

Snow, R. (1942). Further Experiments on Whorled Phyllotaxis, *New Phytol.*, **41**, 108–124.

Snow, R. (1948). A New Theory of Leaf Formation, *Nature*, **162**, 798.

Snow, R. (1949). A New Theory of Phyllotaxis, *Nature*, **163**, 332.

Snow, R. (1951). Experiments on Bijugate Apices, *Phil. Trans. Roy. Soc. London*, **235B**, 291–310.

Snow, R. (1955). Problems of Phyllotaxis and Leaf Determination, *Endeavour*, **14**, 190–199.

Snow, R. (1958). Phyllotaxis of *Kniphofia* and *Lilium candidum*, *New Phytol.*, **57**, 160–167.

Soberon, J. M., and Delrio, M. C. (1981). The Dynamics of a Plant-Pollinator Interaction, *J. Theor. Biol.*, **91**, 363–378.

Steeves, T. A., and Sussex, I. M. (1972). *Patterns in Plant Development*, Prentice-Hall, N.J.

Sterling, C. (1945). Growth and Vascular Development in the Shoot Apex of *Sequoia sempervirens* (Lamb) Endl. II. Vascular Development in Relation to Phyllotaxis, *Am. J. Bot.*, **32**, 380–386.

Stevens, P. S. (1978). *Les Formes dans la Nature*, Seuil, Paris.

Steward, F. C. (1968). *Growth and Organization in Plants*, Addison-Wesley, London.

Stewart, B. M. (1966). *Theory of Numbers*, 2nd ed., Macmillan, New York.

Stewart, W. N. (1964). An Upward Outlook in Plant Morphology, *Phytomorphology*, **14**, 207–209.

Stoll, R. R. (1952). *Linear Algebra and Matrix Theory*, McGraw-Hill, New York.

Sweet, S. S. (1980). Allometric Inference in Morphology, *Am. Zool.*, **20**, 643–652.

Sykes, Z. M. (1969). On Discrete Stable Population Theory, *Biometrics*, **25**, 285–293.

Szabo, Z. (1930). A dipsacaceák virágzatának fejlödestani értelmezese [Entwicklungsgeschichtliche Deutung des Blütenstandes der Dipsacaceen]. *A Szent István Akadémia mennyiségtan-tereszettudományi osztályának felolvasásai*, **2**, 3–72.

Tait, P. G. (1872). On Phyllotaxis, *Proc. R. Soc. Edinb.*, **7**, 391–394.

Taylor, A. (1967). *Introduction to Functional Analysis*, Wiley, New York.

Thom, R. (1975). *Structural Stability and Morphogenesis*, Benjamin, Reading, Mass.

Thomas, R. L. (1975). Orthostichy, Parastichy and Plastochrone Ratio in a Central Theory of Phyllotaxis, *Ann. Bot.*, **39**, 455–489.

Thomas, R. L., and Cannell, M. R. (1980). The Generative Spiral in Phyllotaxis Theory, *Ann. Bot.*, **45**, 237–249.

Thomas R. L., Ng, S. C., and Wong, C. C. (1970). Phyllotaxis in the Oil Palm: Arrangement of Male-Female Florets along the Spikelets, *Ann. Bot.*, **34**, 107–115.

Thompson, D'Arcy W. (1917, 1942, 1968). *On Growth and Form*, 2 vols., Cambridge University Press, Cambridge. Abridged ed., 1971.

Thornley, J. H. M. (1975a). Phyllotaxis I: A Mechanistic Model, *Ann. Bot.*, **39**, 491–507.

Thornley, J. H. M. (1975b). Phyllotaxis II: A Description in Terms of Intersecting Logarithmic Spirals, *Ann. Bot.*, **39**, 509–524.

Thornley, J. H. M. (1976). *Mathematical Models in Plant Physiology*, Academic Press, New York.

Thornley, J. H. M. (1977). A Model of Apical Bifurcation Applicable to Trees and Other Organisms, *J. Theor. Biol.*, **64**, 165–176.

Thrall, R. M., Mortimer, J. A., Rebman, K. R., and Baum, R. F., Eds. (1967). *Some Mathematical Models in Biology*, Rep. #40241-R-7, The University of Michigan Press.

Tomlinson, P. B., and Wheat, D. M. (1973). Bijugate Phyllotaxis in Rhizophoreae, *Bot. J. Linn. Soc.*, **78**, 317–321.

Tort, M. (1969). Modifications Phyllotaxiques Provoquées par des Traitements Thermiques et par l'Acide Gibbérellique chez le Crosne du Japon, *Stachys sieboldii* Miq., *Mem. Bull. Soc. Bot. Fr.*, 179–195.

Trucco, E. (1956a). A Note on the Information Content of Graphs, *Bull. Math. Biophys.*, **18**, 129–135.

Trucco, E. (1956b). On the Information Content of Graphs: Compound Symbols; Different States for Each Point, *Bull. Math. Biophys.*, **18**, 237–253.

Truemper, K. (1978). Algebraic Characterizations of Unimodular Matrices, *SIAM J. Appl. Math.*, **35**, 328–332.

Tucker, S. C. (1961). Phyllotaxis and Vascular Organization of the Carples in *Michelia fuscata*, *Am. J. Bot.*, **48**, 60–71.

Tucker, S. C. (1962). Ontogeny and Phyllotaxis of the Terminal Vegetative Shoots of *Michelia fuscata*, *Am. J. Bot.*, **49**, 722–737.

Turing, A. M. (1952). The Chemical Basis of Morphogenesis, *Phil. Trans. R. Soc. London*, **237B**, 37–72.

Varga, R. S. (1962). *Matrix Iterative Analysis*, Prentice-Hall, *Englewood Cliffs, N.J.*

Veen, A. H. (1973). A Computer Model for Phyllotaxis Based on Diffusion of an Inhibitor on a Cylindrical Surface, part of a thesis presented to the Moore School of Electrical Engineering, University of Pennsylvania, Philadelphia.

Veen, A. H., and Lindenmayer, A. (1977). Diffusion Mechanisms for Phyllotaxis, Theoretical, Physico-Chemical and Computer Study, *Plant. Physiol.*, **60**, 127–139.

Vieth, J. (1964). Le Capitule de Dipsacus Représente-t-il un Système Bijugué? *Mem. Bull. Soc. Bot. Fr.*, 38–47.

Vieth, J. (1965). Etude Morphologique et Anatomique de Morphoses Induites par Voie Chimique sur Quelques Dipsacacées, D.Sc. Thesis, Faculté des Sciences of the Université de Dijon.

Vinogradov, I. M. (1954). *Elements of Number Theory*, Dover, New York.

Vitanyi, P. M. B. (1980). Lindenmayer Systems: Structure, Languages and Growth Functions, *Math. Center Tracts*, Amsterdam.

Voeller, B. R., and Cutter, E. G. (1959). Experimental and Analytical Studies of Pteridophytes. XXXVIII. Some Observations in Spiral and Bijugate Phyllotaxis in *Dryopteris aristata* Druce, *Ann. Bot. N.S.*, **23**, 391–396.

Vogel, H. (1979). A Better Way to Construct the Sunflower Head, *Math. Biosci.*, **44**, 179–189.

Vogel, T. (1956). *Physique Mathématique Classique*, Colin, Paris.

Waddington, C. H., Ed. (1968–1972). *Towards a Theoretical Biology*, Edinburgh University Press, Edinburgh, pp. 1–4.

Wardlaw, C. W. (1949a). Leaf Formation and Phyllotaxis in *Dryopteris aristata*, *Ann. Bot. N.S.*, **13**, 164–198.

Wardlaw, C. W. (1949b). Phyllotaxis and Organogenesis in Ferns, *Nature*, **164**, 167–169.

Wardlaw, C. W. (1949c). Experiments on Organogenesis in Ferns, *Growth* (suppl.), **13**, 93–131.

Wardlaw, C. W. (1952). *Phylogeny and Morphogenesis*, Macmillan, London.

Wardlaw, C. W. (1955). Evidence relating to the diffusion-reaction theory of morphogenesis, *New Phytol.*, **54**, 39–48.

Wardlaw, C. W. (1956). The Inception of Leaf Primordia, in *The Growth of Leaves Proc.*, F. L. Milthorpe, Ed., Butterworths, London, pp. 53–65.

Wardlaw, C. W. (1957). On the Organization and Reactivity of the Shoot Apex in Vascular Plants, *Am. J. Bot.*, **44**, 176–185.

Wardlaw, C. W. (1965a). The Organization of the Shoot Apex, in *Encyclopedia of Plant Physiology*. Vol. 1, W. Ruhland, Ed., Springer-Verlag, New York, pp. 424–442, 443–451, 966–1076.

Wardlaw, C. W. (1965b). *Organization and Evolution in Plants*, Longman, London.

Wardlaw, C. W. (1968a). *Morphogenesis in Plants*, Methuen, London.

Wardlaw, C. W. (1968b). *Essays on Form in Plants*, Manchester University Press, Manchester.

Wardlaw, C. W., and Cutter, E. G. (1956). Experimental and Analytical Studies of Pteridophytes XXXI. The Effect of Shallow Incisions on Organogenesis in *Dryopteris aristata* Druce, *Ann. Bot.*, **20**, 39–57.

Wareing, P. F., and Phillips, I. D. J. (1978). *The Control of Growth and Differentiation in Plants*, Pergamon Press, New York.

Waterman, T. H. (1968). Systems Theory and Biology—View of a Biologist, in *Systems Theory and Biology*, M. D. Mesarovic, Ed., Springer-Verlag, New York, pp. 1–37.

Waterman, T. H., and Morowitz, H. J., (Eds. (1965). *Theoretical and Mathematical Biology*, Blaisdell, New York.

Weisse, A. (1904). Untersuchungen über die Blattstellung an Cacteen und anderen Stamin-Succulenten, *J. Wiss. Bot.*, **34**, 343–423.

Wells, C. (1976). Some Applications of the Wreath Product Construction, *Am. Math. Mon.*, **83**, 317–338.

Wetmore, R. H. (1943). Leaf Stem Relationship in the Vascular Plants, *Torreya*, **43**, 16–28.

Wetmore, R. H. (1956). Growth and Development in the Shoot System of Plants, *Soc. Dev. Growth, Symp.*, **14**, 173–190.

Weyl, H. (1952). *Symmetry*, Princeton University Press; Princeton, N.J. (translated under the title *Symétrie et Mathématique Moderne*, Flammarion, Paris, 1964).

Whaley, G. W. (1961). Growth as a General Process, in *Encyclopedia of Plant Physiology*, Vol. 14, W. Ruhland, Ed., Springer-Verlag, New York, pp. 71–112.

White, A. T. (1973). *Graphs, Groups and Surface*, North-Holland, Amsterdam.

Whyte, L. L., Wilson, A. G., and Wilson, D., Eds. (1969). *Hierarchical Structures*, American Elsevier, New York.

Wiesner, J. (1907). *Der Lichtgenuss der Pflanzen*, Engelmann, Leipzig.

Williams, R. F. (1975). *The Shoot Apex and Leaf Growth: A Study in Quantitative Biology*, Cambridge University Press, Cambridge.

Wilson, C. L. (1953). The Telome Theory, *Bot. Rev.*, **19**, 417–437.

Wright, C. (1873). On the Uses and Origin of the Arrangements of Leaves in Plants, *Mem. Am. Acad. Arts Sci.*, **9**, 379–415.

Young, D. A. (1978). On the Diffusion Theory of Phyllotaxis, *J. Theor. Biol.*, **71**, 421–432.

Zimmermann, W. (1953). Main Results of the Telome Theory, *Paleobotanist*, **1**, 456–470.

SUBJECT INDEX

*The problem of phyllotaxis
holds an important place
in plant morphogenesis, and
botanometry is one of the
most living chapters of
modern science.*

AUTHOR INDEX

Jean, R.V., IX, XII, XIII, XIV, 7, 8, 9, 17, 19, 21, 22, 56, 57, 60, 63, 64, 65, 85, 86, 89, 92, 115, 142, 143, 144, 146, 150, 154, 156, 159, 161, 162, 171, 172, 173, 182
Jensen, L.C.W., 159

Kappraff, J., 60, 162
Khintchine, A.I., 55, 56, 161
Kilmer, W.L., 56
King, C.E., 170
Klein, F., 7, 8, 13, 15, 19, 20, 56, 161, 174
Knuth, D.E., 60
Kuich, W., 160

Laki, K., 57
Laplace, P.S., 102, 107
Larson, P.R., 159
Lehninger, A.L., 94, 101, 161, 170
Leigh, E.G., 57
Leonardo da Vinci, 2, 173
Leppik, E.E., 170
Leslie, J., 154
Leveque, W.S., 56
Levin, D.A., 170
Lie, M.S., 161
Lindenmayer, A., 48, 108, 110, 111, 115, 135, 136, 137, 138, 157, 166
Linnaeus, C. von, VII
Loiseau, J.E., XIII, 59, 91, 138, 163, 164
Lucas, E., 5

McCulloch, W.S., 56, 142
McCully, M.E., 166
Majumdar, G.P., 59, 150
Maksymowych, R., 74, 84, 85, 91, 92, 94
Marzec, C., 60, 162
Mathai, A.M., 89, 90, 172
Maynard-Smith, J., 112
Meicenheimer, R.D., 91
Meinhardt, H., 94, 114
Mendel, G., VII
Mitchison, G.H., 51, 54, 108, 117, 154, 155
Mowshowitz, A., 160

Nägeli, C., 159
Namboodiri, K.K., 159
Novalis, F., 115

O'Connell, D.T., 2

Papentin, F., 160
Perron, O., 154
Phidias, 3, 178
Philipson, W.R., 159, 164
Pilet, P.E., 161
Planck, M., 102, 178
Plantefol, L., XIV, 66, 164, 175
Popper, K., 172
Prigogine, I., 160
Pythagoras, 2

Rashevski, N., 96, 100, 160
Reinberg, A., 146
Rényi, A., 145
Richards, F.J., XI, XIII, 31, 46, 59, 63, 64, 66, 71, 77, 78, 79, 80, 81, 83, 87, 89, 90, 91, 94, 116, 117, 138, 157, 160, 161, 164, 165, 166, 168, 169, 170, 171, 172, 173, 174
Richter, P.H., 157, 172
Ridley, J.N., 89, 161
Roberts, D.W., 59, 156, 157, 168
Rosen, R., IX, XII, XVII, 145, 173
Rutishauser, R., 59

Sachs, J., 65
Schimper, C.F., 5, 14, 65, 76, 159, 173, 178
Schoute, J.C., XIV, 5, 23, 59, 66, 94, 95, 161, 164, 168, 173, 182
Schranner, R., 157, 172
Schrödinger, E., 145, 160
Schwabe, W.W., 59, 64, 89, 91, 164, 170
Schwendener, S., XIV, 66, 118, 122, 123, 156, 164, 167, 173
Shannon, C.E., 160
Sharp, W.E., 141
Sinnott, E.W., 59, 164, 170
Smith, B.W., 175
Smolukovski, 93, 99, 102
Snow, M., and Snow, R., XIV, 66, 91, 123, 124, 126, 129, 135, 138, 159, 161, 164, 166, 167, 169, 173
Steeves, T.A., 165
Steiner, J., 161
Steinhaus, H., 60
Sterling, C., 157, 159
Stevens, P.S., 61
Steward, F.C., 92, 159, 160
Stewart, B.M., 122
Stewart, W.N., 159, 172, 173